International High-Technology Competition

International High-Technology Competition

F. M. Scherer

Harvard University Press
Cambridge, Massachusetts
London, England   1992

Library of Congress Cataloging-in-Publication Data

Scherer, F. M. (Frederic M.)
    International high-technology competition / F. M. Scherer.
      p. cm.
    Includes index.
    ISBN 0-674-45845-1 (alk. paper)
    1. High technology industries—United States—Management.
2. Competition, International.   3. Comparative advantage
(International trade) I. Title.
HD62.37.S34 1992
338.6'048—dc20
91-44479
CIP

Contents

Acknowledgments

The research reported in this book would not have been possible without the generous support and counsel of many organizations and individuals.

The statistical research in Chapter 4 and especially Chapter 5 was carried out in collaboration with Keun Huh, now at the Samsung Economic Research Institute in Korea.

Full-time work by the author and co-investigator Huh at the United States Bureau of the Census during the first two-thirds of 1990 was supported under an American Statistical Association/Census research fellowship funded by National Science Foundation grant SES 87-13643. The materials have been reviewed to ensure that confidential Census data are not disclosed. Additional financial support came from an O'Melveny and Myers Centennial research grant. The findings are those of the author and do not necessarily reflect the views of the National Science Foundation, the Census Bureau, the American Statistical Association, or O'Melveny and Myers.

Computer expertise, data access, and much other invaluable support were provided at the Census Bureau by Arnold Reznek and Robert McGuckin. Assistance and counsel in accessing and interpreting Census records also came from Census staff members Steve Andrews, Elinor Champion, Andy Mary, and Harvey Monk and from Margaret Grucza of the National Science Foundation.

The Chapter 3 case studies were researched by Judy Chevalier, Andreas Beckmann, Richard Gross, John Lonnquist, Anne McCormick, and Michele Rivard. Judy Chevalier and Matthew Barmack served as research assistants on diverse other matters. Susan Herrin helped prepare the manuscript for publication.

Many insights came from critical reactions in seminar presentations at the Census Bureau, the National Bureau of Economic Research; the John F. Kennedy School, Graduate School of Business Administration, and Department of Economics at Harvard University; the University of Maryland; the International Trade Commission; Trinity College in Dublin; the University of Cambridge; and the Science Centre Berlin. Particularly valuable comments were received from Zvi Griliches, Richard Rosenbloom, David Yoffie, Richard Caves, Raymond Vernon, and Dennis Mueller. The author is also grateful for critical comments from referees for Harvard University Press, the *Review of Economics and Statistics,* and *Research Policy.* Portions of Chapter 5 are drawn with permission from F. M. Scherer and Keun Huh, "Top Managers' Education and R&D Investment," *Research Policy* (1992) and F. M. Scherer and Keun Huh, "R&D Reactions to High-Technology Import Competition," *Review of Economics and Statistics* 74.

Important data sets were provided by Larry Katz and John Abowd through work supported by the National Bureau of Economic Research, and William Sullivan of the International Trade Administration, U.S. Department of Commerce. Helpful advice on data sources came from Robert Stern, Richard Caves, and Geza Feketekuty.

Several of the case studies were read by industry executives who played an active role in the events described. Their critical comments and corrections enabled significant improvements.

Seldom has it been more true that without the generous help of the persons and organizations named above, the research presented here would not have been possible. Responsibility for errors of commission and omission remains with the author.

November 1991

International High-Technology Competition

1 Introduction

During the 1960s, the world was mesmerized by the pervasive dominance of American technology. In a widely cited book, the French publisher Jean-Jacques Servan-Schreiber characterized the U.S. lead in industrial research and development investment as "overwhelming" and warned that "if we [the Europeans] fail to catch up," by 1980 "the Americans will have a monopoly on know-how, science, and power."[1] The Organisation for Economic Co-Operation and Development (OECD) commissioned a series of "technology gap" studies, whose main focus was the gap between U.S. technology and that of other industrialized nations.[2]

The situation changed dramatically during the ensuing two decades. By 1980, it was clear that America was far from having a monopoly on industrial science and technology. Other nations did catch up. Indeed, they did more than catch up. Their industries moved visibly ahead in automobile design and quality, steel-making technology, consumer electronic goods, shipbuilding, textile machinery, general-purpose numerically controlled machine tools, copying machines, ceramics, high-voltage electrical transmission and rectifying apparatus, nuclear power reactors, industrial enzymes, and much else.

It was probably inevitable that a single nation could not continuously dominate the world's technological frontiers, given the desire of others to close the gap and make their own innovative contributions. Yet there

---

1. *The American Challenge* (New York: Atheneum, 1968), pp. 63, 101 (translated from the 1967 French edition, *Le Défi Américain,* by Ronald Steel).

2. See OECD, *General Report: Gaps in Technology* (Paris, 1968), p. 16, which found that U.S. companies originated 60 percent of the significant innovations tabulated since 1945 and received 50 to 60 percent of the patent and technology license fees.

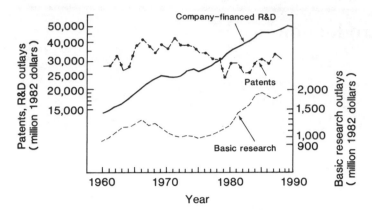

Figure 1.1   Trends in U.S. industrial R&D outlays, basic research, and
patents, 1960–1988. Data from U.S. National Science Board,
Science and Engineering Indicators—1989 (Washington, D.C.:
U.S. GPO, 1989), pp. 271 and 351; and U.S. Bureau of the
Census, *Statistical Abstract of the United States* (various years,
for patent data).

is more to the story of the 1970s and 1980s than relentless technological
convergence.[3] American industry also faltered perceptibly. Figure 1.1
presents three summary measures of U.S. industrial innovative activity,
in each of which there are signs of diminishing vitality. Company-financed
industrial research and development expenditures (solid line) grew in real
(constant 1982 dollar) terms at a fairly constant 6.4 percent annual rate
during the 1960s—when the American challenge seemed so formidable.
But in 1970, they broke precedent by declining, and in the next year,
they fell again before resuming a slower, more erratic upward trend. Brisk
growth resumed in the late 1970s, but since 1986, renewed retardation
is evident. Had constant-dollar R&D continued to grow in the 1970s
and 1980s at the 6.4 percent annual rate sustained during the 1960s—
a counter-factual which, to be sure, could not be maintained indefinitely[4]—1988 R&D outlays would have been 70 percent greater than

3. Compare William J. Baumol, Sue Anne Batey Blackman, and Edward N. Wolff,
Productivity and American Leadership (Cambridge: MIT Press, 1989), chap. 5.

4. See Derek J. de Solla Price, *Little Science, Big Science* (New York: Columbia
University Press, 1963), p. 19, who observes that if worldwide scientific effort continued
to grow at the rates observed over three centuries, in less than one more century "[w]e
should have two scientists for every man, woman, child, and dog in the population."

those actually recorded. The number of U.S. invention patents issued to U.S. corporations (dot-dash line in Figure 1.1), which had been growing at an average rate of 4.3 percent during the 1950s and 1960s, peaked at 43,022 in 1971 and then declined to less than 30,000 per year on average in the 1980s. Even before the fall in industrial R&D and patenting became evident, constant-dollar industry-financed and -conducted basic research (broken line) peaked in 1966, declined sharply, and failed to surpass its 1966 level again until 1981.

Although other influences may have had a more significant overall causal impact, one consequence of diminished growth in research, development, and innovation was a lower rate of productivity growth—that is, most simply measured, the growth of goods and services output per hour of work effort.[5] Productivity growth by this measure for the U.S. private business sector averaged 2.72 percent per year in the 1950s, 2.81 percent in the 1960s, 1.25 percent in the 1970s, and 1.07 percent in the 1980s.[6] The persistent decline of productivity growth from earlier post–World War II attainments has done considerable damage to the American dream—that is, the expectation that each new generation of citizens enjoys material living standards substantially improved over those of the preceding generation.[7]

The technological virtuosity of national industries has been found also to influence in important ways comparative advantage in international trade, and hence the composition of export and import flows. A more detailed exploration of the theory and evidence must await Chapters 2 and 4. For our present purposes, Figure 1.2 tells the relevant story. Manufactured goods are divided into two categories—high-technology products and "all others." The high-technology goods are those that embody relatively intensive research and development inputs, either directly at the final manufacturing stage or through the intermediate goods used in their production.[8] A prominent common feature for both groups is the

---

5. My own estimates, which take into account more inter-industry technology spillovers than those of other scholars and hence find higher social rates of return on R&D investment, suggest that the retardation of R&D growth during the 1970s reduced productivity growth by 0.2 to 0.4 percentage points per annum. F. M. Scherer, "R&D and Declining Productivity Growth," *American Economic Review,* 73 (May 1983), pp. 215–218.

6. *Economic Report of the President* (Washington, D.C.: U.S. GPO, February 1992), p. 348.

7. See "What Happened to the American Dream?" *Business Week,* August 19, 1991, pp. 80–84.

8. On the methodology, see U.S. Department of Commerce, International Trade Ad-

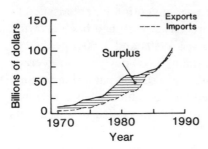

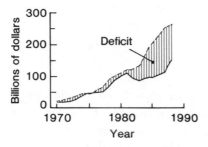

Figure 1.2   U.S. trade balances in high-technology industries *(top panel)* and all other manufacturing industries *(bottom panel)*, 1970–1988. Data from U.S. National Science Foundation, *Science and Technology Data Book: 1990* (Washington, D.C., June 1990), p. 38.

sharp deterioration of U.S. trade balances between 1981 and 1986, largely as a result of the unsustainably high value of the U.S. dollar relative to other foreign currencies during much of that period. High-technology goods differed in sustaining a positive—until 1981, a growing positive—trade balance, whereas the balance for lower-technology manufactures was almost persistently adverse. But whatever comparative advantage U.S. manufacturers possessed in high-technology products, it was not sufficient to resist the macroeconomic forces of the 1980s. The balance of

ministration, *Technology Intensity of U.S., Canadian, and Japanese Manufactures, Output, and Exports* (Washington, D.C.: June 1988). The principal industry categories are synthetic resins and plastic materials; pharmaceuticals; industrial inorganic chemicals; ordnance; engines and turbines; computers and accounting equipment; communications equipment; electronic components; aircraft, guided missiles, and spacecraft; and scientific and professional instruments.

high-technology trade plunged and turned negative for one year (1986).[9] A more detailed breakdown reveals that throughout the 1980s, the U.S. balance of high-technology trade was *negative* with respect to Japan and the newly industrializing nations of East Asia while remaining robustly positive for other developed and less-developed nations.[10] These quantitative data mirror the evidence from qualitative observation that the most potent, though by no means the sole, challenges to U.S. technological leadership have come from Japan.[11] Details will be marshalled in Chapter 3.

How U.S. manufacturing enterprises responded to the high-technology challenges of foreign competitors—in effect, the "new kids on the block"—and how they came into a position of being challenged are the principal questions addressed in this volume. We proceed from the fundamental premise that international comparative advantage in the production and sale of high-technology goods is not something obtained and sustained by historical birthright. Rather, it must be struggled for and earned through superior technological innovativeness. In the jargon of economists, high-technology comparative advantage is an *endogenous* variable—one whose value is determined by the interplay of competitive market processes, broadly construed. If U.S. firms are to maintain an advantage in high-technology trade, it must first and foremost be because they excel at developing superior new products and the production methods needed to bring these products into the marketplace. Or to extend the analysis in an important way, producers may seek advantage on cer-

---

9. An alternative analysis finds that even though the United States lost ground in the balance of high-technology trade, U.S.-based multinational corporations maintained a roughly constant share of the world's high-technology trade—that is, by increasing their high-technology exports from other nations. Irving B. Kravis and Robert E. Lipsey, "Technological Characteristics of Industries and the Competitiveness of the United States and Its Multinationals," National Bureau of Economic Research working paper no. 2933 (April 1989).

10. See U.S. National Science Board, *Science and Engineering Indicators: 1989* (Washington, D.C., 1989), p. 379.

11. In a statistical analysis, Audretsch and Yamawaki find that by 1977, Japan established particularly favorable trade balances with the United States in industries of high R&D intensity. David B. Audretsch and Hideki Yamawaki, "R&D Rivalry, Industrial Policy, and U.S.-Japanese Trade," *Review of Economics and Statistics*, 70 (August 1988), pp. 438–447. See also Robert F. Owen, "The Evolution in Japan's Relative Technological Competitiveness since the 1960s," *Bank of Japan Monetary and Economic Studies*, 6 (November 1988), pp. 75–128.

tain more mature goods (including agricultural products) by achieving relatively low costs through the use of advanced process technology.

Underlying our analysis of high-technology rivalry in manufacturing will be our implied answer to a more fundamental question: Why should one care whether comparative advantage comes from superior technology, as contrasted, say, to superior human capital (which is a necessary concomitant to superior technology), abundant fertile land, or the accumulation of physical capital?

On this question no clear consensus exists among scholars.[12] A partial answer is that, absent strategic control of some scarce natural resource such as petroleum, high material standards of living can be enjoyed by a nation's inhabitants only when advanced technologies are mastered. The story of growing material prosperity in industrialized nations during the past three centuries is essentially an account of implementing continuously improved technologies.[13] This much is uncontroversial, even if commonly ignored in economists' abstract theorizing.

More controversial is the question, does national prosperity depend upon *relative* mastery across the whole array of industrial technologies, or does it suffice that technological leadership is widely shared, with trade among technologically proficient but specialized nations advancing the common good? Here the lack of conclusive theory and evidence may require a conclusion that widespread proficiency plus trade may suffice.[14]

---

12. For various views, see Laura D. Tyson, "Managed Trade: Making the Best of the Second Best," and the accompanying discussion in Robert Z. Lawrence and Charles L. Schultze, *An American Trade Strategy: Options for the 1990s* (Washington, D.C.: Brookings, 1990); Rachel McCulloch, "The Challenge to U.S. Leadership in High Technology Industries: Can the United States Maintain Its Lead? Should It Try?" in Günter Heiduk and Kozo Yamamura, eds., *Technological Competition and Interdependence* (Seattle: University of Washington Press, 1991), pp. 192–211; Jagdish Bhagwati, *The World Trading System at Risk* (Princeton: Princeton University Press, 1991), chap. 3; Michael Porter, *The Competitive Advantage of Nations* (New York: Free Press, 1990), chaps. 1–4; and Paul R. Krugman, *Rethinking International Trade* (Cambridge: MIT Press, 1990), chaps. 9–11.

13. See Joseph A. Schumpeter, *Capitalism, Socialism, and Democracy* (New York: Harper and Row, 1942), especially chaps. 5 and 7; David S. Landes, *The Unbound Prometheus* (Cambridge: Cambridge University Press, 1969); and Joel Mokyr, *The Lever of Riches* (Cambridge: Cambridge University Press, 1990).

14. See Daniele Archibugi and Mario Pianta, "Specialization and Size of Technological Activities in Industrial Countries," in Mark Perlman and F. M. Scherer, eds., *Entrepreneurship, Technological Innovation, and Economic Growth* (Ann Arbor: University of Michigan Press, 1992). They show that specialization on a narrow range of technologies is greater, the smaller a nation is. Or to put the point the other way around, only the largest nations diversify their efforts across the whole array of technologies.

But two doubts intrude. First, the strength of a nation's position in advanced industrial technology can affect the terms of trade, even if not the sum of benefits enjoyed collectively by trading partners, and permit the technically superior partner to capture a disproportionate share of those benefits.[15] Second, widespread mastery of advanced industrial technology may generate certain benefits external to the individual firms implementing the technologies, that is, benefits not appropriated by those firms, but captured instead by others, and disproportionately, by residents of the technology-originating nations rather than by their trading partners. The examples of this proposition derived from static economic models on this point are unpersuasive. More compelling may be the dynamic benefits, as the conquest of one technology leads over time to proficiency in ever more powerful evolutions during subsequent generations, accelerating technical progress, and hence the ability to wrest more from nature, in a kind of virtuous spiral.[16] Failure to maintain the technological pace, on the other hand, can undermine a nation's ability to implement future advances, inducing a vicious spiral of industrial stagnation like that experienced by Great Britain during much of the twentieth century.[17] Clearly, the "British sickness" entailed more than merely falling behind other nations in the implementation of advanced technologies. But that failure was an important component, warning us that widespread loss of technological vitality can have profoundly adverse long-run consequences.

The case is not proved, however, and it may not be proved until it is too late to recoup, so we advance to the burden of our analysis. We seek to illuminate how, and how successfully, U.S. industrial enterprises have responded to high-technology challenges from abroad. We begin in Chap-

---

15. This was clearly recognized by Alfred Marshall, the leading British economist of a century ago: "[I]n the eighteenth century . . . England had something approaching to a monopoly of the new methods of manufacture; and each bale of her goods would be sold—at all events when their supply was artificially limited—in return for a vast amount of the produce of foreign countries . . . But . . . her improvements have been followed, and latterly often anticipated, by America and Germany and other countries: and her special products have lost nearly all their monopoly value." *Principles of Economics* (eighth ed.; London: Macmillan, 1920), pp. 672–674.

16. See Porter, *The Competitive Advantage,* chap. 4; and W. Brian Arthur, "Positive Feedbacks in the Economy," *Scientific American,* February 1990, pp. 92–99.

17. See the contributions by Bernard Elbaum and William Lazonick, Julia Wrigley, and David C. Mowery, in Elbaum and Lazonick, eds., *The Decline of the British Economy* (Oxford: Clarendon, 1986), pp. 1–17, 162–188, and 189–222.

ter 2 by laying out the essential economic theory—notably, on how technological virtuosity affects comparative advantage and the structure of international trade and on how rivalry affects individual firms' efforts to achieve technological innovations. In Chapter 3, we present eleven historical case studies of U.S. industries subjected to significant high-technology challenges from foreign rivals. Chapter 4 continues the empirical analysis in quantitative form, investigating how the detailed structure of U.S. manufactured goods trade was affected by research and development spending and a battery of other technological and structural variables. Chapter 5 then taps detailed U.S. Census data to illuminate how the R&D spending of 308 companies reacted to generally rising high-technology import competition. Chapter 6 pulls the threads together and considers policy implications.

2   Innovation, Comparative Advantage, and R&D Competition

Theories of Comparative Advantage

How comparative advantage leads to gains through international trade was first stated clearly by David Ricardo at the onset of the nineteenth century.[1] Why one nation should use relatively less labor producing some outputs than did other nations was left slightly vague. Important contributions by Eli Heckscher and Bertil Ohlin directed attention to the relative abundance of inputs and biases in input-output relationships.[2] They showed that, in a world of internationally immobile inputs, nations that possess a relative abundance of capital will specialize in producing for export commodities that require a relatively capital-intensive input mix, importing the labor-intensive goods they demand from nations with abundant labor but little capital (for instance, less-developed countries).[3]

---

1. David Ricardo, *On the Principles of Political Economy and Taxation* (London, 1817), chap. 7.

2. Eli F. Heckscher, "The Effect of Foreign Trade on the Distribution of Income" (1919), translated in American Economic Association, *Readings in the Theory of International Trade* (Philadelphia: Blakiston, 1949), pp. 272–300; and Bertil Ohlin, *Interregional and International Trade* (Cambridge: Harvard University Press, 1933). Ohlin's qualitative observations anticipated many later developments—for instance, on labor skills (pp. 83–86), product differentiation and intra-industry trade (pp. 94–96), economies of scale (pp. 52–58 and 106–108), and process innovation (pp. 517–523).

3. In the late 1940s, economists employed by the Central Bank of Japan, following Heckscher-Ohlin logic, argued that Japan, like any less-developed country, should emphasize the production for export of products made by its abundant, low-wage labor. Ministry of Trade and Industry (MITI) staff argued successfully that Japan should instead seek to develop goods embodying advanced technology. See Christopher Freeman, *Technology Policy and Economic Performance: Lessons from Japan* (London: Pinter, 1987), pp. 34–35.

Much theoretical work extending the basic Heckscher-Ohlin paradigm followed.

A significant challenge to the so-called factor abundance view came in a paper by Wassily Leontief in 1953.[4] At the time, the United States was uniquely well endowed among the world's nations in industrial capital, while its labor was relatively scarce and wages were correspondingly high. From this, Leontief reasoned, the United States should export goods whose production was relatively capital-intensive and import labor-intensive goods. Analyzing 1947 data, he found the opposite—a labor-intensive bias. Although later work reconciled Leontief's results with traditional factor abundance theory,[5] the "Leontief Paradox" took the economics profession by storm. As a critic observed much later, "It is . . . difficult to find another empirical result that has had as great an impact on the intellectual development of the discipline."[6]

One hypothesis advanced to rationalize the Leontief Paradox was that American labor was itself capital-intensive, embodying rich "human capital" from education and on-the-job skill development. Thus, U.S. firms were exploiting their comparative advantage by exporting skill-intensive commodities. From this it was but a short step to hypotheses contending that the United States' comparative advantage lay in products that were research and development–intensive.[7]

Statistical Evidence

While such hypotheses, to be explored further in a moment, were proliferating, empirical support came from two papers published simultane-

---

4. Wassily Leontief, "Domestic Production and Foreign Trade: The American Capital Position Reconsidered," *Proceedings of the American Philosophical Society,* 97 (September 1953), pp. 332–349.

5. See Edward E. Leamer, "The Leontief Paradox, Reconsidered," *Journal of Political Economy,* 88 (June 1980), pp. 495–503; and *Sources of International Comparative Advantage* (Cambridge: MIT Press, 1984), chap. 2.

6. Edward E. Leamer, "Leontief Paradox," *The New Palgrave: A Dictionary of Economics,* vol. 3 (London: Macmillan, 1987), p. 166.

7. For more extensive surveys of hypotheses formulated during this period, see Giovanni Dosi, Keith Pavitt, and Luc Soete, *The Economics of Technical Change and International Trade* (London: Harvester Wheatsheaf, 1990), chap. 2; Stephen P. Magee, *International Trade* (Reading, Mass.: Addison-Wesley, 1980), chap. 3; and Edward M. Graham, "Technological Innovation and the Dynamics of the U.S. Comparative Advantage in International Trade," in Christopher T. Hill and James M. Utterback, eds., *Technological Innovation for a Dynamic Economy* (New York: Pergamon, 1979), pp. 122–124.

ously in a leading economics journal.[8] They revealed that, when the manufacturing sector was disaggregated to between 18 and 22 industry groups, the groups that were the most R&D-intensive accounted for a disproportionate share of U.S. exports. Also, there was a high correlation between tallies of scientists and engineers as a proportion of the American work force and U.S. industries' share of exports originating from ten leading nations. Both papers found that other variables, such as labor skills and a propensity toward large-scale industrial operations, played a complementary role. Subsequent research using much richer data confirmed the importance of technological innovation as a key component of U.S. industries' export success. Four variables measuring the R&D or innovative content of exports outranked all but two of 31 alternative variables (measuring capital intensity, labor skills, metallurgical inputs, and much else) in importance as inputs into 354 industries' net balance of exports over imports for 1967.[9] The only higher-ranked inputs were coal and cropland (which explained the large net exports of the coal mining and agricultural sectors). Broadening the investigation to 22 industrialized nations, Giovanni Dosi and colleagues found that differences in national innovative effort, measured by the share of 1963–1977 patent grants received by national companies in the United States, were by far the leading determinant of manufactured goods trade balances. The more technically innovative a national industry was, the more exports from that industry tended to exceed imports.[10]

Since the time when most of these analyses were completed, more finely disaggregated data on industrial research and development spending have become available through the "Line of Business" surveys conducted for four years by the U.S. Federal Trade Commission.[11] As a

8. William Gruber, Dileep Mehta, and Raymond Vernon, "The R&D Factor in International Trade and International Investment of United States Industries," and Donald B. Keesing, "The Impact of Research and Development on United States Trade," *Journal of Political Economy*, 75 (February 1967), pp. 20–48.

9. Leo Sveikauskas, "Science and Technology in United States Foreign Trade," *Economic Journal*, 93 (September 1983), pp. 542–554.

10. Dosi, Pavitt, and Soete, *The Economics of Technical Change and International Trade*, pp. 167–185.

11. Federal Trade Commission, Bureau of Economics, *Statistical Report: Annual Line of Business Report, 1977* (Washington, D.C., April 1985). The report for 1977 was both the last and most comprehensive of the four Line of Business surveys—hence its use in all of the correlations here. Analyses of more aggregated data have shown that industry R&D/sales ratios are highly correlated over periods as long as 15 years. See F. M. Scherer, *Innovation and Growth: Schumpeterian Perspectives* (Cambridge: MIT Press, 1984), p. 273.

prelude to work that will be reported later, 1977 R&D/sales ratios were matched with ratios of exports, imports, and net exports (that is, exports minus imports) to domestic output value for 449 U.S. manufacturing industries. Table 2.1 reports the simple (Pearsonian) correlations between R&D/sales ratios by industry and the three measures of U.S. international trade. Because imports (and net exports) were divided by the dollar value of domestic output, rather than by domestic consumption, it is possible for imports/sales and net exports/sales figures to exceed 100 percent, sometimes by a wide margin, in industries whose consumers were supplied more by imports than by domestic production. Because such high "outlier" values of the trade performance variable can swamp the results covering other industries, Table 2.1 also presents correlations for a subsample from which 23 industries with net exports/sales percentages exceeding 100 are deleted. For both data sets, the R&D-import correlations are all close to zero. Mirroring the results first obtained by William Gruber and colleagues at a much higher degree of aggregation for the

Table 2.1   Correlations of trade indices with industry R&D/sales ratios

Year	All 449 industries			23 industries excluded[a]		
	Imports/ sales	Exports/ sales	Net exports/ sales	Imports/ sales	Exports/ sales	Net exports/ sales
1965	−.008	.335	.155	−.015	.323	.259
1970	.023	.383	.150	−.003	.386	.277
1975	.012	.369	.167	.023	.376	.281
1980	.010	.370	.125	.038	.385	.265
1981	−.001	.379	.118	.034	.386	.255
1982	−.017	.383	.097	.041	.395	.240
1983	−.017	.388	.084[b]	.031	.403	.229
1984	−.016	.368	.053[b]	.034	.384	.178
1985	−.009	.350	.043[b]	.022	.363	.180
1986	−.009	.338	.039[b]	.032	.350	.154

a. The excluded industries (with SIC codes in parentheses) are miscellaneous carpets and rugs (2279), lace goods (2292), miscellaneous textile products (2299), fur goods (2371), raincoats (2385), leather clothing (2386), special sawmill products (2429), rubber and plastic footwear (3021), shoe findings (3131), women's footwear (3144), miscellaneous footwear (3149), women's handbags (3171), fine earthenware products (3263), miscellaneous pottery products (3269), miscellaneous nonferrous metals (3339), secondary smelter products (3341), calculators (3574), radio and television sets (3651), motorcycles and bicycles (3751), watches and clocks (3873), jewelers' findings (3915), dolls (3942), and feathers and artificial flowers (3962).

b. The 0.05 significance level with 449 observations is .093; the 0.01 level is .122.

1960s (see note 8), the R&D-export correlations are consistently positive and statistically significant. More interesting are the correlations between R&D and net exports, that is, with imports deducted from exports to assess the net balance of trade. Though weaker than the gross export correlations, the net export correlations were positive and statistically significant at the 0.01 level from 1965 through 1980. After 1980, however, there was a marked decline, and for the full 449-industry sample, the correlations drop below the 0.05 statistical significance level after 1982. Evidently the general upsurge of imports during the early 1980s (attributable primarily to macroeconomic influences) was especially pronounced in R&D-intensive industries that had previously been strong net exporters.

The Product Cycle Theory

One of the most richly developed explanations of observed U.S. trade patterns is the product cycle (or life cycle) theory, propounded by Raymond Vernon in 1966.[12] According to Vernon, the United States provided a particularly propitious climate for the pioneering of new products because it was the largest and most prosperous nation, ensuring strong demand-pull, especially for new products with a high income elasticity of demand. Also, its large market afforded greater ability to absorb heavy R&D costs and the output from large-scale production facilities. Once such products were developed and marketed, latent but weaker demands for them in other national markets were satisfied, at first through exports from the locus of innovation. Hence the correlation between R&D intensity, which is much more closely associated with product than process innovation,[13] and net export balances. As new products moved into later stages of their life cycles, however, important changes occurred. The innovating firm's desire to hurdle import barriers in other industrialized nations, to reduce product shipping costs, to satisfy differentiated national product design demands, and to solidify localized preferences led to the establishment of offshore production operations. Because the inno-

---

12. "International Investment and International Trade in the Product Cycle," *Quarterly Journal of Economics*, 80 (May 1966), pp. 190–207.

13. For 449 four-digit manufacturing industries, the simple correlation between company-financed R&D/sales ratios and the fraction of R&D expenditures associated with internal production process improvements in 1974 was −0.31. The data are adapted from Scherer, *Innovation and Growth*, chap. 3.

vating firm often possessed unique production know-how and brand image advantages not easily transferred to other organizations, it was profitable for the innovator to own and control these new operations.

Out of the product life cycle theory emerged an explanation of the multinational corporation (MNC), an organizational form proliferating rapidly from the United States during the 1950s and 1960s. As other companies imitated and perhaps improved upon the new product, price competition became more vigorous and pressure to reduce production and marketing costs increased. At first, decentralizing production into primary market areas was the preferred means of cost reduction. But as the relevant product technology matured further and competition from imitators intensified, producers sought new ways of cutting costs. These frequently involved locating production—now a matter of technological routine—in low-wage nations, characteristically, less-developed countries. At this point, classic Heckscher-Ohlin comparative advantage considerations came to the fore. The operations in less-developed nations might or might not be owned by the originally innovating MNC. In either case, to escape the increasing pressure on profits as its products entered the mature phases of their life cycle, the originally innovating firm was motivated to undertake new product innovations, starting the cycle all over again.

In the years following the product cycle theory's first comprehensive expression, world economic conditions evolved in ways that required a modification of its predictions. First, as companies in diverse parts of the world became adept at the game, competitive imitation lags shortened and the maturation process accelerated. Consequently the period during which the home nation of product innovators enjoyed an export advantage shrank, which could reduce R&D–net export correlations. Second, other industrialized nations experienced more rapid economic growth than did the United States, and as a result the profitability of those nations as a locus for the first innovative phase of a product's life cycle rose. The extension of markets through perfection of the European Common Market and the European Free Trade Association reinforced this tendency. Third, with the spread of the multinational corporation form to companies based in Europe and, at least initially, even without such a structure in Japan, business enterprises came increasingly to view the markets for their products as global rather than national. With the world as their market, firms with home bases in small nations could compete at product innovation on terms virtually equal to those enjoyed by companies residing in large national markets. The market size advantages that had often nominated the United States as a first locus of innovation therefore dwindled in importance.

Paralleling the articulation of the product cycle theory as an explanation for high-technology trade and foreign direct investment was an increasing focus on product life cycle considerations in normative theories of corporate strategy.[14] In early stages of the life cycle, it was argued, rich growth opportunities made it profitable to invest heavily inter alia in research and new product development. But in later stages, as growth rates ebbed, prominent U.S. strategic planning consultants advised that products should be treated as "cash cows"—which meant, among other things, cutting back investment in R&D. Also, in the R&D that remained, the focus shifted from developing or improving products to perfecting production processes. Observed average business patterns fit this set of prescriptions well.[15] In the next chapter we shall see that they are relevant to understanding company behavior in some of our case studies.

Theories of Intra-Industry Trade

One limitation of the product life cycle theory is that it predicts a rather lopsided flow of high-technology products from large nations to small. Yet for both the United States and (especially) the individual nations making up the European Common Market, it was observed that there were appreciable two-way international trade flows within industries with relatively high levels of research, development, and product innovation. In other words, there was substantial *intra-industry trade*.[16] Even before these relationships were statistically verified, Staffan Linder proposed an alternative model relating trade to product innovation.[17] In Linder's schema, the specific attributes consumers seek from products differ ap-

---

14. See, for instance, Bruce D. Henderson, *Henderson on Corporate Strategy* (Cambridge, Mass.: Abt, 1979), especially pp. 82–85 and 163–166; and Michael E. Porter, *Competitive Strategy* (New York: Free Press, 1980), especially chaps. 10–12.

15. See Robert D. Buzzell and Bradley T. Gale, *The PIMS Principles* (New York: Free Press, 1987), pp. 200–204.

16. See especially Bela Balassa, "Tariff Reductions and Trade in Manufactures," *American Economic Review,* 56 (June 1966), pp. 466–473; and Herbert G. Grubel and P. J. Lloyd, "The Empirical Measurement of Intra-Industry Trade," *Economic Record,* 47 (December 1971), pp. 494–517.

17. Staffan Linder, *An Essay on Trade and Transformation* (Stockholm: Almqvist & Wicksell, 1961).

preciably from one nation to another. Historically, U.S. consumers favored large cars and Europeans small cars; West German consumers preferred washing machines with a boiling water pre-soak, while Italians favored less complex designs. Physically differentiated product variants were developed in nations, not always the largest, whose consumer preferences they satisfied best. But consumers within a nation are not homogeneous, and so some possibly sizable minority of consumers in Nation A will prefer the products developed in Nation B, causing exports to flow from B to A. For identical reasons, a flow from A to B emerges. The result is a more or less evenly balanced set of intra-industry trade flows, and the more heterogeneous are consumers' tastes, the more balanced the trade flows.

The essence of the Linder rationale for intra-industry trade is product differentiation, which was also the focus of Edward Chamberlin's path-breaking theory of monopolistic competition.[18] An immediate implication of it (and also of the product cycle theory, at least for early phases) is that sellers enjoy some discretion over pricing. They are in effect monopolists, though limited by the competition of imperfect substitutes and potential entrants. This inference created new challenges for theories of international trade, which had traditionally emphasized pure competition.

Another implication of intra-industry trade theory is at least equally important. When products are physically differentiated and the demand for individual variants is limited owing to the diversity of consumer tastes, economies of scale may be less than fully exhausted. This happens in three main ways.

First, launching a new product requires research, development, and initial marketing costs that are "sunk" once production commences. Although sunk costs are variable before the innovation process begins, once they are incurred they are fixed costs when one analyzes how total costs per unit vary with the volume sold per time period. If (taking a long-run

---

18. Edward H. Chamberlin, *The Theory of Monopolistic Competition* (Cambridge: Harvard University Press, 1933). When products are differentiated, defining the boundaries of an "industry" and hence measuring "intra-industry" trade become more difficult. See, for example, Robert Triffin, *Monopolistic Competition and General Equilibrium Theory* (Cambridge: Harvard University Press, 1960), chap. 2; and F. M. Scherer and David Ross, *Industrial Market Structure and Economic Performance* (third ed.; Boston: Houghton-Mifflin, 1990), pp. 73–79 and 176–184. Some economists have argued that these difficulties preclude empirical research on intra-industry trade, but their view is overly pessimistic, at least for the fine level of industry disaggregation at which we work in this volume.

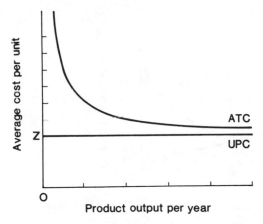

Figure 2.1 Product-specific scale economies with front-end fixed costs

post-innovation perspective) variable unit production costs are the same per unit across all plausible output volumes, the average *total* cost curve becomes a rectangular hyperbola, as illustrated in Figure 2.1. Average production costs, *UPC,* are assumed constant at *OZ* per unit. When front-end R&D and marketing costs (appropriately capitalized, as under mortgage financing[19]) are prorated over the output volume, a continuously declining average total cost curve, *ATC,* is obtained. Economies of scale prevail at all annual output levels.

Second, especially when product differentiation has proceeded to the point at which each product variant captures only a modest share of total demand for the relevant class of products, there are likely to be product-specific economies of scale. These stem from fixed machine setup costs, the necessity for job shop (as contrasted to automated) production methods, and the like.[20] Thus, even ignoring front-end R&D

---

19. If *RD* is the initial outlay on R&D, incurred at time zero, if the economic life of the resulting product will be *T* years, and if the appropriate interest rate is *r*, the annual fixed financial charge on a capitalized R&D mortgage will be:

$$\frac{RD}{\dfrac{1 - (1 + r)^{-T}}{r}}.$$

20. See F. M. Scherer, Alan Beckenstein, Erich Kaufer, and R. D. Murphy, *The Economics of Multi-Plant Production: An International Comparisons Study* (Cambridge: Harvard University Press, 1975), pp. 49–56, 295–321, and 355–381. The increasingly widespread adoption of flexible (e.g., computer-driven) manufacturing methods is making this source of scale economies progressively less important.

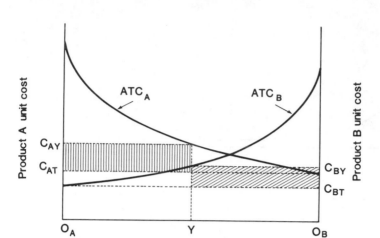

Figure 2.2   Gains from trade with product-specific scale economies

costs, the long-run average cost curve will be downward-sloping until substantial output volumes are attained. The advantages of international trade in this case (and by extension in the Figure 2.1 case) are illustrated in Figure 2.2. The cost curve for Product A is ATC_A, drawn conventionally to show persistent (though diminishing) economies of scale as output increases (moving from left to right). Product B's cost curve is drawn unconventionally, with its zero point O_B on the right-hand side of the diagram, and with output rising from right to left. (The axis scales calibrating the output of A and B need not be identical.) Suppose now that minimum unit costs on these two cost functions will be attained only if production is on a sufficient scale O_AO_B to satisfy the demands of *two* equal-sized nations. If each nation 1 and 2 satisfies its demands for A and B autarkically, the *two* plants supplying A will each produce only OY units, at average total cost O_AC_{AY}, which is appreciably higher than the unit cost O_AC_{AT} that could be attained if all output were concentrated at one location and exports satisfied the other nation's demand for A. The excess cost from forgoing specialization and international trade is two times the vertically shaded rectangle (less unspecified transportation costs). Similarly, autarkic production of B forgoes product-specific scale economies and leads to unit costs higher than necessary by twice the diagonally shaded rectangle. If, to achieve all attainable scale economies, Nation 1 specializes in producing A and Nation 2 in producing B, the resulting intra-industry trade will yield substantial cost savings.

An important analogue of product-specific scale economies comes from the phenomenon known loosely as "learning by doing."[21] In assembly operations, workers become more dexterous and devise cost-reducing shortcut methods as they accumulate experience on the job. In delicate tasks such as integrated circuit production, operators learn through repeated experience how to control the process parameters, reducing time losses between batches and, more crucially, improving good "chip" yields from one or two per hundred at early stages to as much as 90 percent when substantial experience has been accumulated. Characteristically, the relationship between batch unit costs and cumulative output volume (not output per time period, as in conventional cost curves like those in Figures 2.1 and 2.2, but the accumulation of total output over time) has been found to be roughly linear on doubly logarithmic coordinates. Figure 2.3 provides an illustration for the early years of monolithic integrated circuit production. Assuming that prices actually charged provide a reasonable approximation of how costs fell with experience, dramatic cost reductions are implied. The "slope" of the fitted solid straight-line "learning curve" is 77.6 percent, which means that with each cumulative doubling or redoubling of output, unit circuit costs fell by (100 − 77.6) = 22.4 percent—a fairly typical experience.

That costs can be reduced so much through learning by doing again implies substantial benefits from intra-industry trade, with a firm in one nation producing all of one relevant product variant and supplying the world while other firms, perhaps located in other nations, similarly specialize and export. Learning-by-doing economies are also conducive to intriguing price rivalry dynamics. The firm that races down the learning curve first, for example, to a cumulative output of 200,000 "chips" in Figure 2.3, enjoys a cost advantage of roughly $3 per unit over the firm that has produced, say, only 20,000 chips (with unit costs of roughly $5). Thus an advantage accrues to the "first mover." That advantage can be reinforced by pricing aggressively to accelerate user adoption of the new product, thereby enhancing cumulative volume, and to win the lion's

---

21. See Harold Asher, *Cost-Quantity Relationships in the Airframe Industry* (Santa Monica: RAND Corporation study R-291, July 1956); Armen Alchian, "Costs and Output," in Moses Abramovitz et al., eds., *The Allocation of Economic Resources* (Stanford: Stanford University Press, 1959), pp. 23–40; Jack Hirshleifer, "The Firm's Cost Function: A Successful Reconstruction?" *Journal of Business*, 35 (July 1962), pp. 235–255; and Kenneth J. Arrow, "The Economic Implications of Learning by Doing," *Review of Economic Studies*, 29 (April 1962), pp. 155–173.

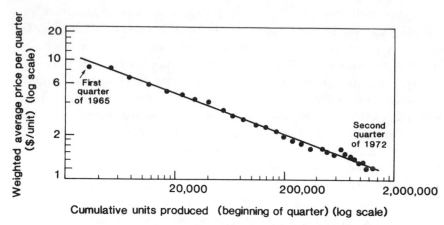

Figure 2.3   Learning curve for monolithic integrated circuits, 1965–1972.
Adapted with permission from Boston Consulting Group, Inc.,
Perspectives, No. 124, "The Experience Curve—Reviewed, I.
The Concept," 1974.

share of that volume from emerging competitors. With a sufficiently aggressive pricing policy, early innovators can maintain dominant, if not monopoly, market shares for an extended period and, among other things, sustain substantial profits once a solid cost advantage has been established and competitors are held in check.[22]

Two main conditions limit the extent to which learning by doing facilitates and (for minimum costs) necessitates monopoly production. First, learning curves eventually flatten out—a phenomenon that helps signal the onset of the mature phase of the product life cycle.[23] Second, not all the cost-reducing benefits of learning by doing are appropriated by the

---

22. See A. Michael Spence, "The Learning Curve and Competition," *Bell Journal of Economics,* 12 (Spring 1981), pp. 49–70; Drew Fudenberg and Jean Tirole, "Learning-by-Doing and Market Performance," *Bell Journal of Economics,* 14 (Autumn 1983), pp. 522–530; Chaim Fershtman and Uriel Spiegel, "Monopoly versus Competition: The Learning by Doing Case," *European Economic Review,* 23 (November 1983), pp. 217–222; and David R. Ross, "Learning to Dominate," *Journal of Industrial Economics,* 34 (June 1986), pp. 337–353.

23. The evidence on this is sparse. But see Asher, *Cost-Quantity Relationships,* chaps. 4 and 7. There is also weak evidence that competition leads to steeper learning curves than would be found under monopoly conditions. See F. M. Scherer, *The Weapons Acquisition Process: Economic Incentives* (Boston: Harvard Business School Division of Research, 1964), pp. 119–126.

firms that carry out the early production. Some knowledge of cost-reducing methods "spills over" to actual and potential competitors, domestic and foreign, as others "reverse engineer" the resulting product, hire away key personnel, make plant inspections (if permitted), read process patent specifications, and pay for know-how licenses.[24] Little is known about the degree to which such spillovers occur, but to the extent that they do, laggard rivals are less disadvantaged, and the compulsions toward monopoly product market structure are weakened.

The "New" International Trade Theory

The sellers of differentiated products in early product life cycle stages enjoy some degree of monopoly power. Substantial economies of scale, static or dynamic, support pricing strategies that entrench or defend monopoly power and that contribute, at minimum, to oligopolistic market structures—that is, markets in which only a few sellers vie for position. Recognition of this led, beginning in the late 1970s, to the emergence of a "new" international trade theory (NITT).[25] The new analyses emphasize monopolistic competition, monopoly pricing, economies of scale,[26] and the exploration of strategies through which individual firms and their host nations can achieve product market dominance and skew the terms of trade in their favor. With their concern for monopoly and oligopoly strategy, they draw from the same theoretical core as the field of specialization in economics known as "industrial organization," which traditionally has focused on the functioning of national rather than international markets. As the NITT literature began to blossom, work on game-theoretic indus-

---

24. See, for example, Ross, "Learning to Dominate," pp. 345–348; and Pankaj Ghemawat and A. Michael Spence, "Learning Curve Spillovers and Market Performance," *Quarterly Journal of Economics,* 100 (Supplement 1985), pp. 839–852.

25. Surveys and collections of pioneering papers include Avinash K. Dixit and Victor Norman, *Theory of International Trade* (Cambridge: Cambridge University Press, 1980); Elhanan Helpman and Paul R. Krugman, *Market Structure and Foreign Trade: Increasing Returns, Imperfect Competition, and the International Economy* (Cambridge: MIT Press, 1985); Paul Krugman, ed., *Strategic Trade Policy and the New International Economics* (Cambridge: MIT Press, 1986); and Paul Krugman, *Rethinking International Trade* (Cambridge: MIT Press, 1990).

26. The NITT literature usually refers to "increasing returns to scale" rather than "economies of scale." Increasing returns to scale refers to production function input-output relationships, economies of scale to unit cost-volume relations. The relationships share a mathematical duality, but not in simply characterized ways for realistic cases. In our view, economies of scale is the more operational concept.

trial organization models was exploding. The two fields of specialization have therefore shared similar methodologies, advanced in parallel, and indeed had a mutually reinforcing impact.

Much research of high quality has been accomplished in the NITT framework. An appreciable fraction is relevant to this book. Yet two limitations must be recognized.

First, the underlying issues have been investigated with more or less success by both industrial organization and international trade specialists for the better part of a century. Advances continue, to be sure, but careful inquiry reveals that our substantive understanding of the relevant phenomena has not changed radically relative to the state of the art in, say, the 1960s. Much of the new work adds mathematical rigor and technical articulation to concepts that have long been known. This is admitted by Paul Krugman, a leading practitioner of NITT:

> Since economics as practiced in the English-speaking world is strongly oriented toward mathematical models, any economic argument that has not been expressed in that form tends to remain invisible. While many economists no doubt understood that increasing returns could explain international trade even in the absence of comparative advantage, before 1980 there were no clean and simple models making the point. As a result this idea was often simply left out of textbooks and trade courses . . . [T]here is an influential body of informal literature on technology and trade . . . This does seem to appeal to practical men but has not been stated in the form of models and is therefore still ambiguous in its implications.[27]

Whether mathematical formalism and rigor are essential to the success of theory, or merely a current affectation of the economics profession, is a question we shall not debate here. What is undisputed is that the essential concepts of NITT are "old" in the sense of having been recognized and intuitively understood, but "new" in the sense of being reduced to precise, tractable mathematical models.

Second, the game theory used in most NITT models is an extremely flexible instrument. With some small perturbation of assumptions, almost any behavioral outcome can be generated. But this is a serious shortcoming. A theory that can predict everything explains nothing.[28] It has come

---

27. Krugman, *Rethinking International Trade*, pp. 3 and 153.
28. See also Franklin Fisher's review, "Organizing Industrial Organization: Reflections on the *Handbook of Industrial Organization*," in *Brookings Papers on Economic Activity*,

to be recognized that merely proliferating differentiated theories of firms' strategic interactions adds little to our understanding of real-world behavior unless it is accompanied by parallel empirical tests.[29] This too is acknowledged by Krugman: "At this point . . . the central problem of international trade is how to go beyond the proliferation of models to some kind of new synthesis. Probably trade theory will never be as unified as it was a decade ago, but it would be desirable to see empirical work begin to narrow the range of things that we regard as plausible outcomes."[30]

The present work attempts to honor that admonition. Ignoring the fashion of the times, it proliferates no new theories of international trade or industrial organization. Those we have in abundance already. Rather, it seeks to illuminate empirically how real-world business enterprises interact in an environment of international competition characterized by technological innovation, product differentiation, economies of scale, and the strategic quest for competitive advantage.

Models of Innovation Rivalry

We are not yet done with theory, however. The new international trade theory operates for the most part on a lofty plane of abstraction, with faceless national champions vying for strategic position in the world arena. Little assimilated into the trade literature is a substantial body of analysis, mostly by industrial organization economists, on how individual firms or groups of firms square off to compete in the development and commercialization of new products and production processes. Here, too, we have suffered a surfeit of riches. A late 1980s survey covering only the innovation timing literature lists 75 theoretical contributions.[31] As in other domains where game theory and similar models of inter-firm rivalry are used, the results are highly sensitive to the assumptions, and with the

1991, *Microeconomics*, pp. 201–225. His "organizing principle 2" states: "The principal result of theory is to show that nearly anything can happen."

29. For analogies to the methodology of physical science, see F. M. Scherer, "On the Current State of Knowledge in Industrial Organization," in H. W. de Jong and W. G. Shepherd, eds., *Mainstreams in Industrial Organization* (Dordrecht: Kluwer, 1986), pp. 5–22.

30. Krugman, *Rethinking International Trade*, p. 261.

31. Jennifer F. Reinganum, "The Timing of Innovation: Research, Development, and Diffusion," in Richard Schmalensee and Robert D. Willig, eds., *Handbook of Industrial Organization*, vol. 1 (Amsterdam: North-Holland, 1989), pp. 850–908.

appropriate constellation of assumptions, virtually anything can happen. This is not terribly helpful as a guide to case studies and statistical investigations, so we focus here mainly on the simplest and (one might argue) most fundamental insights.

Some Basic Propositions

We begin with several propositions on the "production" of technological innovations, which, already oversimplifying, we assume takes place in the context of company R&D programs:

P1. As time passes, the cost of carrying out a project to develop an end product or production process with given performance characteristics falls, sometimes discontinuously, owing to the accumulation of facilitating scientific and technological knowledge.

P2. From a given starting date and hence a given stock of facilitating knowledge, it is possible to pursue "fast" or "slow" R&D schedules, but within limits, a faster pace of development can be achieved only by incurring higher R&D cost.

P3. Research and development entails substantial uncertainties. They decline as R&D progresses.

P4. Technical uncertainties can be hedged by pursuing multiple conceptual approaches in parallel. Parallel (as opposed to serial) exploration of alternative concepts reduces expected time to successful project completion, but raises expected R&D costs. This is one basis for the time-cost trade-off relationship of (P2). Parallel developments can be carried out either within a single firm or across a set of competing or cooperating firms.

P5. For a given starting date and hence a given knowledge stock, the greater the improvement relative to existing product or process performance characteristics, the more R&D costs rise. In other words, extra "quality" costs more.[32]

P6. Differences in the "efficiency" of organizations in carrying out R&D may be sufficiently great as to swamp the cost relationships implied in (P1), (P5), and especially (P2).

---

32. For the first known formalization of this relationship, see M. J. Peck and F. M. Scherer, *The Weapons Acquisition Process: An Economic Analysis* (Boston: Harvard Business School Division of Research, 1962), p. 469. Propositions (P1) through (P4) are also conceptualized in Chapters 9 and 10 of that book.

These propositions cover the "supply" side of the technological innovation process. The "demand" side is more complex and much more sensitive to alternative assumptions. We begin with three most elementary propositions:

D1. For given assumptions about demand, the vantage point in time, market structure, rival behavior, and production and capital costs, one can define a function showing a would-be innovator's expected quasi-rents (sales revenues less production and distribution costs) from marketing a new product, as they depend upon the timing of product introduction. Perceived from a given R&D project starting date, the more rapidly the project is completed, the larger is the discounted present value of the firm's expected quasi-rents.

D2. When latent demand for a new product is growing exogenously over time, the quasi-rent function identified in (D1) shifts outward as the date of R&D project initiation is delayed.

D3. Innovations made profitable mainly by the growth of demand, leading to the outward quasi-rent function shifts described in (D2), are called "demand-pull" innovations. Innovations made profitable by a fall in the cost of R&D (under P1) are called "technology-push" innovations. Actual inducement mechanisms commonly entail a combination of demand-pull and technology-push stimuli.

Competition versus Monopoly

The simplest case is illustrated in Figure 2.4.[33] We assume that the cost of developing a new product declines over time with the general advance of knowledge. The question is, when to seize the opportunity and carry out the innovation? When the R&D is conducted, we assume to simplify the exposition that it is financed by a mortgage whose annual carrying costs are $C^*(T)$. Because R&D costs decline over time, so does the annual carrying cost $C^*(T)$, as shown in Figure 2.4. Once the innovation is completed, the innovator has a monopoly in its sale and realizes a surplus of revenues over production and distribution costs (that is, quasi-rents)

33. It is adapted from Yoram Barzel, "Optimal Timing of Innovations," *Review of Economics and Statistics*, 50 (August 1968), pp. 348–355.

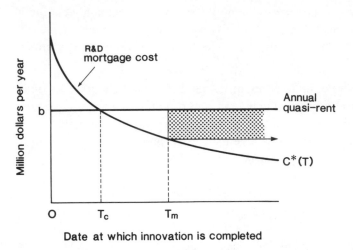

Figure 2.4   Innovation dates under competition and monopoly

of *b* dollars per year, assumed constant over time. This implies that there is no demand-pull effect.

Under unfettered competition for opportunities to innovate and become a product monopolist, expected economic profits are zero. In the present case, this means that competition forces acceleration of the product innovation date until annual R&D mortgage costs $C^*(T)$ are equal to expected annual quasi-rents, at T_c in Figure 2.4. A firm with a secure monopoly position assured *before* innovation occurs would wait longer, however, until R&D mortgage carrying costs have fallen below expected annual quasi-rents, for instance, at time T_m. This ensures a stream of maximum discounted profits from T_m on,[34] given by the dotted rectangular

---

34. The proof is taken from Scherer and Ross, *Industrial Market Structure,* pp. 639–641. Let *b* be the annual quasi-rent realized after innovation occurs, *r* the rate of interest, C_O the one-time cost of development at time $t = 0$, and *p* the rate at which development costs fall over time. Then the cost of development at time *T* is $C_O e^{-pT}$. A monopolist maximizes the discounted present value of quasi-rents less costs:

$$\pi_M = \int_T^\infty b e^{-rt}\, dt - C_O e^{-(p+r)T}.$$

The definite integral is:

$$\frac{b e^{-rT}}{r} - C_O e^{-(p+r)T}.$$

area in Figure 2.4. Thus, under competition as defined starkly here, innovation occurs earlier, and R&D costs are greater, than under monopoly.

Exactly how competition forces the pace to a zero-profit equilibrium is the subject of further assumptions. Under one formulation proposed by Glenn Loury, firms contract irrevocably for R&D projects with uncertain completion dates but a fixed cost commitment.[35] This is in effect a lottery, the discounted present value of whose prize equals the total cost of all entrants' R&D. One firm, favored by chance with the shortest time to development success, wins the prize. Loury shows that given his assumptions, an increase in the number of competing firms causes each individual firm to reduce its own R&D outlay commitment, but the R&D outlays of all firms together increase. Under a variant explored by Tom Lee and Louis Wilde, unsuccessful competitors can discontinue their R&D and cut their losses when a winner emerges.[36] With this small change in the assumptions, it turns out that an increase in the number of competitors leads each firm to increase its annual R&D spending rate. Both aggregate and individual firm R&D outlays then increase with the number of rivals.

Oligopoly Interactions

Technological innovation is more likely to be characterized by imperfect information, recognition lags, and disequilibrium than by the primitive but stringent competitive assumptions specified thus far, under which the

---

Differentiating with respect to T, the first-order condition for maximum profits is:

$$be^{-rT} = (p + c)C_0 e^{-(p+r)T}.$$

By way of contrast, competitive break-even implies setting the profits integral above equal to zero and rearranging to:

$$\frac{be^{-rT}}{r} = C_0 e^{-(p+r)T}.$$

Multiplying both sides by r, we obtain a condition analogous to the monopoly first-order profit maximization equation above:

$$be^{-rT} = rC_0 e^{-(p+r)T},$$

except that the premultiplier on the right-hand side is r rather than $(p + r)$. Competitive break-even must therefore occur at a lower value of T than monopoly profit maximization.

35. Glenn C. Loury, "Market Structure and Innovation," *Quarterly Journal of Economics,* 93 (August 1979), pp. 395–410.

36. Tom Lee and Louis Wilde, "Market Structure and Innovation: A Reformulation," *Quarterly Journal of Economics,* 94 (March 1980), pp. 429–436. See also the synthesis in Reinganum, "The Timing of Innovation," pp. 855–859.

number of firms and their R&D spending rates adjust without friction to
a zero expected profit equilibrium. Alternative models view the number
of rivals as given by historical or accidental circumstances and then analyze how the addition of more rivals affects R&D spending and the speed
at which new products are introduced.

Crucial to virtually all such analyses is some kind of first-mover advantage. The most common assumption is that the first firm to complete its
R&D successfully receives a patent that allows it to monopolize the relevant product market more or less permanently, subject perhaps to pricing
constraints imposed by non-controlled inferior technologies. Pure cases
of this sort do arise occasionally, as in pharmaceuticals, but it is not
necessary or realistic to accept such extreme assumptions. There is abundant evidence that the first successful innovators tend quite typically to
enjoy enduring reputational and market penetration advantages, manifested in price premiums, more rapid progression down learning curves
and hence lower costs, and larger market shares than those of late innovators or "me too" imitators.[37]

The broadest overview, from a model postulating that the first mover's
retained market share rises as a continuous function of its time lead over
imitators, is provided by Figure 2.5.[38] The horizontal axis measures the
number of years taken from the start of development to product introduction; the vertical axis measures discounted benefits and costs. The R&D
cost curve $C(T)$ shows that, given the state of knowledge at time $T = 0$,
expected R&D costs are higher, the more the time schedule is com-
pressed and, hence, the more closely "crash project" conditions are ap-
proached. Four discounted individual firm quasi-rent curves $V \cdots$ are
presented, depending upon the number of symmetrically positioned rivals. The effect of increasing rivalry is shown by the differing slopes of
the quasi-rent curves. The slopes become steeper with more rivals because the more symmetric rivals there are, the more market share (and
hence profit) a would-be innovator has to lose if it falls behind in the race
to market. Profits are maximized by finding the maximum vertical distance between the quasi-rent and cost curves. For a monopoly (quasi-rent
curve V_M), the profit-maximizing schedule is 4.5 years; for two symmetri-

37. See Scherer and Ross, *Industrial Market Structure*, pp. 430–438, 580–592, and
626–630.

38. See Scherer and Ross, *Industrial Market Structure*, p. 633, drawing upon Scherer,
Innovation and Growth, chaps. 4, 5, and 6.

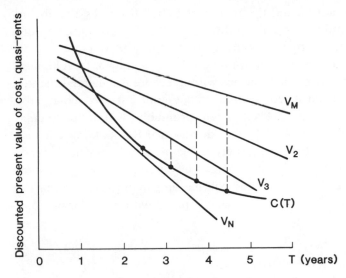

Figure 2.5 How development time depends upon the number of rivals

cally positioned rivals, 3.75 years; and for three rivals, 3.15 years. Greater rivalry stimulates more rapid innovation. But with the curve V_N (for more than three sellers), the situation changes dramatically. *If* the symmetric rivals invest in R&D, they will adopt a still faster 2.5-year schedule. But if each perceives that the innovation race will be a losing proposition, they will invest nothing.

Although this particular result can be avoided if some rivals drop out asymmetrically, it reflects a duality that emerges from many models of R&D rivalry. Up to some point, more rivalry leads to greater R&D expenditures and more rapid project completion. But rivalry that becomes too intense, so that no participant in the race can expect its post-innovation quasi-rents to repay R&D costs, discourages R&D, delays innovation, and may permit no innovation at all. Simply put, the predicted relationship between market structure and the intensity of R&D investment is nonlinear. This "inverted U" hypothesis has been supported in several empirical studies of R&D rivalry within national markets.[39]

Figures 2.4 and 2.5 together have a further implication. When facilitat-

---

39. See Scherer and Ross, *Industrial Market Structure*, pp. 645–647; and William L. Baldwin and John T. Scott, *Market Structure and Technological Change* (Chur: Harwood, 1987), pp. 32–39 and 90–93.

ing knowledge is advancing, the time-cost trade-off curves $V\cdots$ in Figure 2.5 shift toward the origin (southwest) as the date for initiation of a product development project is repeatedly postponed. At some moment in time, $C(T)$ lies below V_M but not below V_2, V_3, and V_N. Development is profitable for a monopolist but not within a more fragmented market. For reasons explained in connection with Figure 2.4, the monopolist, although able to commence development profitably, will prefer to delay its start until $C(T)$ has shifted farther downward. But as knowledge advances, development also becomes profitable in more fragmented market structures. The more rapidly knowledge is advancing and hence the more rapidly $C(T)$ is shifting, the shorter will be the gap between the feasible initial starting dates for oligopolists as compared with those for monopolists.

Reaction Lags and Asymmetries

Because technological rivalry between foreign and domestic firms often entails significant cost and brand recognition differences among the participants or lags in incumbent firms' perception of the threat, we must push further and explore the role of asymmetries.

A useful approach is to examine how firms react to changes in rival R&D time schedules. Concretely, we plot the *Cournot* reaction functions of paired rivals (which may be a group of domestic firms versus a group of foreign firms), postulating that at each step in the time-phased reaction process, Firm 1 chooses the strategy that maximizes its own profit, given the strategy chosen most recently by its rival, Firm 2.

Figure 2.6 introduces the technique for symmetrically positioned rivals.[40] The vantage point in time is the moment when the rivals begin R&D or, if there are recognition lags, when they first become aware of their rivalry. The diagram's axes measure the number of years that will elapse from the moment the rivalry commences until each firm completes its R&D and introduces its new product into the market. Firm 1's time to completion is measured on the horizontal axis, Firm 2's on the vertical axis. If the pair of points representing the two firms' strategy choices falls below the 45° line, Firm 1's product introduction date will be earlier than Firm 2's, and so Firm 1 will be the innovating first mover, reaping

---

40. The technique is adapted from Scherer, *Innovation and Growth*, chap. 5, which reproduces a 1967 article.

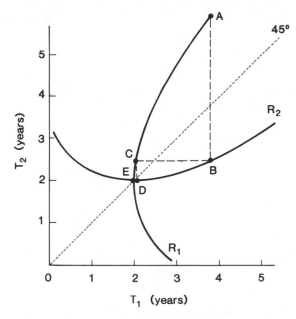

Figure 2.6 Symmetric duopoly R&D schedule reactions

all the advantages a first move entails. If the strategy pair lies above the 45° line, Firm 1 will be the second mover or laggard. R_1 is Firm 1's reaction function, or the locus of profit-maximizing responses to Firm 2's schedule choices. It is inflected at the 45° line where the transition between first- and second-mover status occurs. Firm 2's reaction function, R_2, is analogously defined and, at least in this example, is symmetric to R_1.

Suppose Firm 1 starts the action, choosing a leisurely 3.8 year schedule (point A on R_1) because the would-be innovator does not yet perceive Firm 2's challenge. Firm 2 reacts by moving to B on R_2. Seeing Firm 2 now in the lead, Firm 1 accelerates to C. Firm 2 reacts to D, inducing Firm 1's move to E, which is very near an equilibrium in the sense that no further moves increase either firm's expected profits. Symmetric rivalry has precipitated a series of reactions that accelerated the pace of product innovation to a relatively fast 2.0-year R&D schedule.

Now we inject asymmetries into the picture. Certain companies may be so well entrenched in their markets through customer loyalty or tightly controlled distribution channels that if small or newly entering Firm 2 comes up with its new product months or even years ahead of Firm 1's

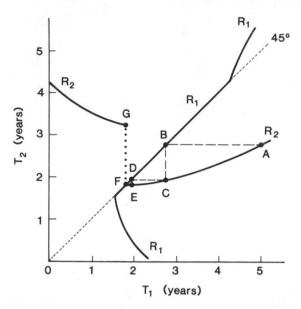

Figure 2.7   Dominant firm reactions (along R_1) to small interloper

introduction date, Firm 1 will hold on to much of its historical market
share. If so, the situation is likely to be described by Figure 2.7. Firm 1
is the entrenched incumbent, Firm 2 the small interloper. Suppose Firm
1, seeing no challenge on the horizon, delays commencing R&D. Firm 2
perceives an opportunity to jump ahead and capture some of what would
otherwise be Firm 1's market share. It chooses point *A* on its reaction
function. But although Firm 1 has little to gain by being ahead (at best,
a larger fraction of what would otherwise be the small market share of
Firm 2), it has much to lose (a chunk of its *large* market share) if it falls
behind.[41] Its reaction function therefore traces the 45° diagonal over an
appreciable range where its position is threatened, so it moves to point
B. Firm 2 accelerates to point *C*, but 1 retaliates to point *D*, and so on
to point *F*. If Firm 2 realizes that it cannot be first, it encounters a reaction
function discontinuity, cutting its R&D costs by retreating to a point like
G. And it may, under some conditions, drop out of the race altogether.
Following the terminology introduced by Lewis F. Richardson in his
pioneering studies of arms races, such reactions by the disadvantaged

---

41. For a geometric illustration, see Scherer and Ross, *Industrial Market Structure*, pp.
635–636.

firm are called "submissive" reactions.[42] Dominant firms tend to be slow starters but to react aggressively when their market shares and profits are threatened by small interlopers. Or in a variant, when there are informational spillovers so that dominant Firm 1 can reduce its own R&D costs by lagging behind Firm 2 and learning from it, dominant firms may adopt a "fast second" strategy. With such a strategy they avoid steps that force the innovative pace, but accelerate their efforts to ensure that their new products lag those of smaller pioneers by no more than a safe, dominance-preserving, interval.[43]

A different scenario unfolds when, for some reason, one firm secures an insuperable lead over the other. The lead may come from a substantial R&D head start unnoticed until too late by the laggard, or a superior R&D organization that permits projects to be carried out more rapidly for given resource inputs,[44] or lower R&D personnel and capital costs. The resulting interactions are illustrated in Figure 2.8. The asymmetry of reaction functions implies a substantial head start for Firm 2, which, absent immediate competition, plans to introduce its new product (at *A*) in year 2.8. When Firm 1 awakes to the challenge, it commences a crash project to recoup, moving to point *B*. Recognizing the competition, Firm 2 accelerates to point *C* (1.4 years). Seeing that it cannot be the first mover, Firm 1 reacts submissively, accepting a solution with an inferior market share but lower R&D costs at *D* (which may trigger further small reactions from Firm 2). Or in rivalries that approximate a winner-take-all solution, Firm 1 may drop out of the race altogether.[45]

Whether the laggard in this case is doomed to inferiority or market exclusion permanently or only temporarily depends upon further assumptions. If the first mover's advantages persist into subsequent product improvement generations, for instance, through reputational ties, persistent learning-by-doing advantages in production, the ability to improve

---

42. Lewis F. Richardson, *Arms and Insecurity* (Chicago: Quadrangle, 1960), chap. 4.

43. See William L. Baldwin and G. L. Childs, "The Fast Second and Rivalry in Research and Development," *Southern Economic Journal*, 36 (July 1969), pp. 18–24; and Reinganum, "The Timing of Innovation," pp. 869–876.

44. There is evidence that Japanese firms have enjoyed such advantages in many developments. See Edwin Mansfield, "The Speed and Cost of Industrial Innovation in Japan and the United States," *Management Science*, 34 (October 1988), pp. 1157–1168; Kim B. Clark, W. Bruce Chew, and Takahiro Fujimoto, "Product Development in the World Auto Industry," *Brookings Papers on Economic Activity*, 1987, no. 3, pp. 729–771; and "What Makes Yoshio Invent?" *The Economist*, January 12, 1991, p. 61.

45. See Reinganum, "The Timing of Innovations," pp. 877–878.

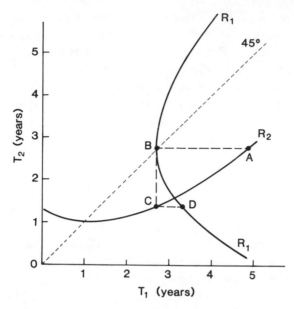

Figure 2.8   Rivalry with head start

existing products at lower cost than others can develop new challengers from scratch, or the avoidance of organizational disruption, firms that lost the initial race and, a fortiori, those that dropped out may have great difficulty recouping in later product generations. But if the participants in new generation rivalries start afresh without significant handicaps or advantages, the firm that finishes a poor second or drops out in one generation may, having learned its lesson, redouble its R&D efforts in the next generation.

Recapitulation

Rivalry in research, development, and innovation can have extremely complex consequences, depending upon the parameters of the rivalry. Although there may be exceptions, the most general theoretical predictions are the following:

1. The introduction of new products tends to be faster, and industry-wide R&D expenditures greater, under competition than under secure monopoly.

2. However, if competition becomes so pervasive and intense that no actor can anticipate positive profits, innovation will be retarded and may fail altogether.
3. When individual oligopolistic firms interact symmetrically in new product rivalries, the most common reaction to intensified rivalry is aggressive. That is, the rate of R&D spending increases and schedules are accelerated.
4. Firms with entrenched advantages that allow them normally to dominate their markets have weak incentives to force the pace of product technology advance, but react aggressively when their positions are threatened by another would-be innovator.
5. Firms caught in an inferior position—for example, because a dominant rival has responded aggressively or because some rival has captured an overwhelming lead—are likely to reduce the intensity of their R&D, and they may drop out of the race altogether. In other words, they react submissively.

All of these generalizations are relevant to the rivalry between U.S. firms and overseas competitors as well as to rivalry within a single national market. However, asymmetric cases (4) and (5) may be especially important in international rivalry. At least during the past two decades, foreign firms challenging U.S. enterprises at the frontiers of technology were very much "new kids on the block," whose capabilities were not immediately recognized and taken into account. Irreversible lags could thus materialize. The overseas challengers were more likely to deviate from accepted norms of the domestic R&D rivalry game, to disrupt existing equilibria, and to adopt organizational forms permitting R&D projects to be completed more rapidly or at lower cost (sometimes with government subsidies). All these imply, at least initially, submissive reactions by challenged domestic incumbents. However, U.S. firms often enjoyed incumbency advantages conducive to "fast second" strategies without serious loss of market share. This suggests the possibility of highly aggressive reactions, especially when large, profitable market positions were at risk. The actual mix of U.S. company reactions remains to be established empirically—the principal task of subsequent chapters.

Differentiated Product Rivalry

Monopolistic competition among new products that are technically differentiated from one another poses analytic challenges that are different in some respects, but parallel in others. To move to the heart of the matter, we adopt a novel expositional approach.[46]

The basic point is that product characteristics can vary more or less continuously along relevant dimensions of an abstraction called "product characteristics space." Automobiles can be small, gargantuan, and all sizes in between; slow or fast in their acceleration; reliable or subject to varying probabilities of breakdown; outfitted with radios using from one to eight loudspeakers; and so on. The properties of engineering resins for making molded plastic parts vary widely in melting point, ultimate tensile strength, surface toughness, opacity, moisture absorbance, resistance to solvents, and so on. Facsimile machines vary in speed, reproduction quality, reliability, operating ease, and so on.

We illustrate the relevant concepts in Figure 2.9 for the case of airliners, assumed (in a severe oversimplification of reality) to vary along a single characteristic, the number of passengers carried in a standard seating configuration.[47] We assume that designing an airliner with particular characteristics is expensive—say, with total R&D costs of $1 billion. At least initially, only three variants are assumed to exist—A, carrying 100 passengers; B, carrying 200; and C, with a capacity of 400. The solid tent-row-shaped line $Q_1 Q_A Q_2 Q_B Q_3 Q_C Q_4$ traces the quasi-rents (that is, the surplus of revenues minus production costs) that can be realized by the airplanes' producer(s). The quasi-rent function has maximum values above characteristics space points A, B, and C, because those designs satisfy ideally the needs of airlines flying routes with such passenger requirements. Planes sold to airlines for routes of other passenger densities yield lower quasi-rents because they satisfy only imperfectly, that is, at higher cost, the requirements of those routes. In effect, they are imperfect substitutes for the needs of those routes. Concretely, in the spirit of our example, quasi-rents are lower in cases entailing imperfect meshing of capacity with demand, because relatively high operating costs and hence fares, or non-optimal scheduling frequencies, choke off traffic on those routes. The less well designs A, B, and C meet the needs of airlines for planes carrying 50, 300, and 500 passengers, the more steeply the

---

46. It is adapted with some simplification from Scherer, *Innovation and Growth*, chap. 8, which reproduces a 1979 paper. See also Scherer and Ross, *Industrial Market Structure*, pp. 602–607.

47. Only graphing difficulties prevent the analysis from being extended to two or N relevant product characteristics, e.g., range, speed, fuel economy, maneuverability, and durability.

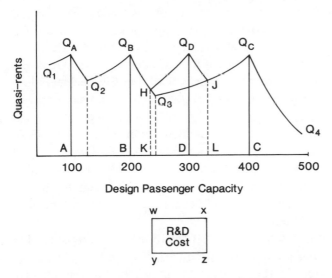

Figure 2.9 Monopolistic competition in airliner development

quasi-rent curves descend from their best-mesh points above *A*, *B*, and *C*. In the present case, planes designed to carry 400 passengers substitute better for hops requiring 300 passengers than they do when 500 passengers must be carried (for instance, by squeezing seating more tightly and restricting range), and so the quasi-rent function slopes downward less steeply to the left of point *C* than to the right.

We add now one more assumption: that the demand for airliners of varying capacities is distributed uniformly over the spectrum of capacities. Thus, if variety had no cost, airlines would buy as many planes optimized for 50 passengers as they would for 250, 300, 350, and so on. This implies that the market is particularly ill served in the neighborhood of 230 to 300 passengers. Suppose now that a firm invests the necessary $1 billion and designs an airliner D optimized to carry 300 passengers. Because the demands in that range are now better served, more planes will be ordered at more attractive prices, thrusting the quasi-rent function upward by the segment HQ_DJ. Airlines will buy plane D for routes requiring 230 to 330 passengers, and aggregate annual quasi-rent realizations will rise by area HQ_DJQ_3.[48]

---

48. Thus, total quasi-rent realizations are the integral under the quasi-rent curve over the airplanes demanded in each size category.

The key question here is, what are the incentives to develop and sell product variant D, and how do they vary with market structure? If a single firm produces aircraft B and C and is reasonably secure in its market position, it will develop D only if its R&D cost (whose magnitude is given graphically by the area of rectangle $WXYZ$ in Figure 2.9) is less than the incremental quasi-rent area HQ_DJQ_3.[49] As the figures are drawn, the opposite is true, and so a "monopolist" of B and C would not wish to develop D. But consider now an "outsider," that is, a monopolistic competitor, with no stake in producing aircraft B and C. Developing D is attractive to the outsider if R&D costs are less than the five-sided quasi-rent area KHQ_DJL. As $WXYZ$ is constructed, this inequality holds, and so the monopolistic competitor will develop D even when the "monopolist" of B and C will not.

The quasi-rent area KHQ_3JL that makes the difference here is surplus that, absent the outsider's entry, the producer of B and C will enjoy, whether or not it develops D. It is in effect "cannibalized" by the outsider from the producer of B and C; hence the asymmetry of incentives. To be sure, if the producer of B and C is persuaded that an outsider will develop D if it fails to do so, the whole area KHQ_DJL becomes forgone quasi-rent to the incumbent, and to avert that loss, the incumbent may preempt the outsider and develop D after all. The need to preempt outsiders is the result of a position at B and C that is insecure against the entry of monopolistic competitors, as distinguished from a position that is secure against the outside entry of physically differentiated products. In this important sense, variations in the character of rivalry affect R&D decisions. Whether the incumbent producer of B and C can avoid competing at *D*, preempts *D*, or is beaten by an outsider depends upon recognition lags, cross-product first-mover advantages, and asymmetries in market access between the incumbent and the outsider.[50] Outside entry at *D* might be particularly likely if the nation in which the outsider enjoys "made at home" preferences has an unusually high incidence of routes requiring 230 to 330 passengers. We shall see that, for the specific capacities postulated, these conditions describe reasonably accurately the early competition between Boeing and Airbus Industrie. The example will also prove applicable to other international rivalry cases.

---

49. Time discounting is ignored for the sake of simplicity.

50. For both the incumbent and the outsider to develop D more or less simultaneously, incurring front-end R&D costs of $1 billion each, would be wasteful.

Normative Ramifications

The overview of R&D rivalry theory presented here has been an exercise in positive economics, attempting to describe how firms behave in diverse structural contexts, not how they *should* behave. But the "should" question must also be raised.

The cannibalized surplus KHQ_3JL in Figure 2.9 that spurs the development of product variant D conveys no gain to society as a whole; it is merely transferred from the producer of B and C to the outsider developing D. As a transfer, it sends a false signal to outsiders (and to preempting insiders), telling them to incur R&D costs without corresponding social gain. Whether those costs *should* be incurred cannot be determined conclusively with the information contained in Figure 2.9. If R&D costs are less than the unambiguous quasi-rent or producer's surplus gain area HQ_DJQ_3, product D *should* be developed (ignoring discount rate complications). But HQ_DJQ_3 is a lower-bound estimate of the social gain from having D. In addition, some addition to consumers' surplus is almost always generated by the availability of more product variants.[51] The magnitude of that surplus is determined by factors similar to those influencing HQ_DJQ_3—the degree to which B and C are good or poor substitutes for D, and (implicit thus far) producers' pricing behavior.

If B and C are poor substitutes, the consumers' surplus gain from having D is likely to be substantial, suggesting that D ought to be developed even though R&D costs exceed HQ_DJQ_3 by modest amounts. As a crude rule of thumb, if R&D costs are less than twice HQ_DJQ_3, the uncounted consumers' surplus may be large enough to justify developing D. When B and C are good substitutes for D, the quasi-rent lines in Figure 2.9 will be fairly flat, HQ_DJQ_3 will be of modest height, and if R&D costs are substantial, development of D is unlikely to be justified.

When the producer of B and C enjoys substantial monopoly power, the prices of B and C may be high relative to production costs and the cannibalizable surplus KHQ_3JL is likely to be large. That large surplus provides powerful incentives for entry by monopolistic competitors with product variant D. A highly competitive situation, in contrast, with prices close to production costs, discourages the entry of differentiated products.

---

51. See especially Edwin Mansfield et al., "Social and Private Rates of Return from Industrial Innovations," *Quarterly Journal of Economics*, 91 (May 1977), pp. 221–240.

To generalize: On the one hand, given relatively high front-end R&D costs, monopolistic competition is more apt to stimulate excessive product variant proliferation, the higher product prices are in relation to production cost and the better substitutes existing products are for the new variants whose introduction is at stake. Monopolistic competition serves society well, on the other hand, when product pricing is relatively (but not purely) competitive, when consumers value product variety highly, and when new product development and launching costs are modest.[52]

The normative balance is similar for other rivalry cases, but more complex. Monopoly with blockaded entry is almost always likely to develop and introduce new products too slowly. The more new products yield non-transferred surplus to consumers as well as producers, the more desirable the accelerating effect of competition is. But the more the competitive race to be first mover adds R&D costs merely to transfer producers' surplus from rivals to oneself, the more likely it is that waste will occur through R&D duplication or excessive "crash project" costs. Vigorous competition can force the pace of innovation too little, too much, or by just the right amount—and it is difficult to be sure which case holds. In product rivalry, unlike the static allocation of resources, no "invisible hand" ensures an optimal result.

Cost-Reducing R&D and Pricing Strategy

In the United States, between one-fourth and one-third of industrial R&D effort is directed toward the development and improvement of companies' own production processes. The lion's share of R&D is oriented toward *product* development and improvement.[53] Product R&D contributes to comparative advantage in international trade primarily through product differentiation.[54] Process R&D, however, is mainly cost-reducing

---

52. For additional generalizations, see Scherer and Ross, *Industrial Market Structure*, p. 606.

53. See Scherer, *Innovation and Growth*, p. 36; and Edwin Mansfield, "Industrial R&D in Japan and the United States: A Comparative Study," *American Economic Review*, 78 (May 1988), pp. 223–228. In Japan, Mansfield's survey suggests, the fraction of R&D devoted to process work is nearer to two-thirds—a striking difference.

54. To be sure, the improved "products" of capital goods and materials suppliers become improved processes for downstream firms. See Scherer, *Innovation and Growth*, pp. 38–51, for measures of inter-industry technology flows. However, to the extent that such products are purchased also by foreign firms, domestic purchasers do not necessarily gain an advantage.

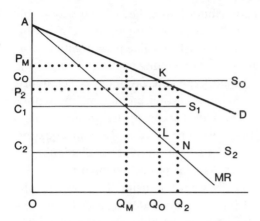

Figure 2.10   Pricing following "run of the mill" and "drastic" process
           innovations

(although quality assurance may also be affected). Cost reduction in turn
influences comparative advantage only if it causes reductions in the innovating industry's prices relative to the prices of all other products supplied domestically. Will it in fact do so?

For polar cases such as pure monopoly with blockaded entry or pure
competition with rapid adoption of the cost-saving innovation by all domestic firms, the answer is clear: prices will fall, although by less than
the amount of the cost reduction over a wide range of supply and demand
conditions. For intermediate cases such as oligopoly, monopoly, or monopolistic competition subject to entry threats, and competition with incomplete diffusion of the innovation, theory yields more ambiguous guidance. From the multitude of possibilities, we pause to examine only two
here.

One, formulated initially by Kenneth Arrow,[55] assumes that before
innovation an industry is competitively structured, with each supplier
producing at the same unit cost, OC_0, in Figure 2.10. Thus the pre-
innovation supply curve is C_0S_0 and the equilibrium price is OC_0. Suppose now that some firm develops an innovation that reduces unit costs
to OC_1. The innovator, assumed to have solid patent protection, can

---

55. Kenneth J. Arrow, "Economic Welfare and the Allocation of Resources for Invention," in the National Bureau of Economic Research conference report, *The Rate and
Direction of Inventive Activity* (Princeton: Princeton University Press, 1962), pp. 619–622.

derive maximum profits from its contribution in either of two ways—by charging a price sufficiently below C_O to drive competitors out of the market, or (more plausibly, if exit is sluggish) by licensing rights to use its innovation to all existing producers and collecting royalties from them. Through its control of output as the ultimate sole producer or its ability to assess royalties, the innovator gains a certain amount of monopoly power. What post-innovation price will maximize its profits? Constructing the conventional marginal revenue curve AMR and equating marginal revenue with cost C_1S_1, we find the conventional monopoly output to be OQ_M and the price OP_M. But a price that high cannot be sustained, because if it were charged, competitors using the inferior technology C_OS_O would undercut it, driving the price down to OC_O and expanding output back to OQ_O. The inability to hold prices above OC_O leaves the innovator with the discontinuous marginal revenue function C_OKLMR. Because the new cost function C_1S_1 cuts that marginal revenue function within its vertical discontinuity KL, the innovator's best attainable policy, implemented either through direct monopoly control or by setting a royalty rate slightly below C_OC_1 per unit produced, is to leave the price at its pre-innovation level, OC_O, and output at its corresponding pre-innovation equilibrium value of OQ_O. If, however, an even greater cost reduction to C_2S_2 is achieved, the new cost function cuts the marginal revenue function at point N, so the monopolist will profitably expand output to OQ_2 and reduce price to OP_2. If this new market equilibrium is reached by licensing competitors, the royalty rate will approximate P_2C_2. Process innovations that induce no expansion of output and hence no price reduction under the assumptions taken here are called "run of the mill" innovations; those that do lead to price reductions and output increases are called "drastic" innovations.[56] A process innovation is more likely to be drastic, the larger the percentage cost reduction it permits, and the more price-elastic the demand for the product whose production it facilitates.

A second case is particularly relevant when a dominant firm, or a group of jointly acting firms, have monopolistic price-setting power in their domestic market but face the threat of steadily growing import competition if the price is set above some "limit price," P_O, equal to potential

---

56. The terminology comes from William D. Nordhaus, *Invention, Growth, and Welfare* (Cambridge: MIT Press, 1969), pp. 70–73.

foreign competitors' unit costs, including inbound freight.[57] The more the actual price $P(t)$ exceeds P_O, the more rapidly imports will flow into the domestic market, capturing a rising market share from the domestic price-setters. If $P(t)$ is set below P_O, imports will decline, and the larger the gap between P_O and $P(t)$, the more rapid the decline will be.

The dominant domestic group's long-run profit-maximizing strategy under these circumstances depends upon four key variables: the discount rate applied to future earnings, the group's domestic market share, the speed at which imports rise or fall with a given gap between $P(t)$ and P_O, and the magnitude of the domestic producers' unit cost advantage or disadvantage relative to overseas rivals' cost P_O. The size of that cost advantage or disadvantage is in turn affected by process innovations. Three main outcomes are possible.

First, if, even after cost-reducing innovation, the domestic group operates either at cost parity to, or with higher unit costs than, overseas rivals, long-run group profits will be maximized by setting the price $P(t)$ above P_O and letting import penetration rise. Process innovation in this case will only moderate the extent to which prices are elevated, slowing but not halting the rate of import growth.

Second, over some range of post-innovation unit costs below P_O, the domestic group will maximize long-run profits by setting its prices in the neighborhood of P_O, holding importers' domestic market share roughly constant. This strategy, called "asymptotic limit pricing," yields a more or less steady stream of supra-normal profits for the domestic group.

Third, if unit costs are reduced sufficiently below importers' delivered cost P_O, the domestic group may set prices below P_O and progressively reduce the importers' domestic market share. For a given domestic firm cost advantage, this "exclusionary" pricing strategy is more likely, the larger importers' pre-innovation market share is, the more rapidly importers exit in response to loss-imposing prices, and the lower is the rate at which domestic firms discount future as compared to current profits. While importers' market share is being driven down, the domestic group sacrifices short-run profits relative to what they would be under a less

---

57. The analysis here is an adaptation of Darius W. Gaskins, "Dynamic Limit Pricing: Optimal Pricing under Threat of Entry," *Journal of Economic Theory*, 3 (September 1971), pp. 306–322. For extensions, qualifications, and illustrations, see Scherer and Ross, *Industrial Market Structure*, pp. 361–396.

exclusionary policy. Eventually, when imports are reduced to some non-zero value, the domestic group will raise prices to P_O.

Thus, cost-reducing innovations may lead to quite different pricing strategies and hence international trade patterns, depending upon complex but identifiable cost and structural relationships. Not all process innovations trigger price reductions, but the likelihood of import-inhibiting price cuts is higher, the more costs are reduced.

Conclusion

During the 1970s and 1980s, U.S. firms have been subjected to technological challenges from new and unfamiliar foreign competitors. The patterns of international trade that have followed these challenges depend critically upon how the U.S. firms have responded in their own R&D efforts. Economic theory provides rich, but in the end inconclusive, guidance concerning likely behavioral responses in such high-technology competitions. To proceed further, empirical evidence is needed.

3 Case Study Evidence on R&D Reactions

The theory of structural R&D rivalry yields ambiguous predictions. An incumbent firm may react to a new competitor's innovative efforts by intensifying its own R&D, by cutting back its R&D or even ceding the relevant market entirely, or by ignoring the threat and continuing business as usual. To understand how domestic companies respond to international high-technology competition, we must observe actual behavior. In this chapter we begin the effort with an analysis of eleven product rivalry case studies.

Several questions receive special attention. What was the origin of the technological challenge from foreign rivals? Why were U.S. companies caught off guard, if indeed they were? How did they initially perceive the threat? What were their initial reactions in terms of R&D project support, pacing, spending levels, and related production and marketing strategy facets? Why were those strategies chosen? What action–reaction patterns followed as the domestic and foreign rivals squared off to compete? Who gained market position, and who lost? Who ended up with the leading position in the new technology's market, and why?

The case studies, some narrow and some broad, include (1) wet shaving apparatus; (2) consumer electronic apparatus, specifically, television sets and video recorders; (3) radial automobile and truck tires; (4) electronic calculators; (5) jet airliners; (6) amateur photographic cameras and film; (7) medical diagnostic imaging equipment; (8) digital central office telephone switches; (9) optical fiber message transmission cable; (10) facsimile machines; and (11) heavy-duty earth-moving equipment. The cases were chosen to illustrate a wide array of technologies, markets, and import challenge outcomes. In all, the four-digit industries to which the relevant products are assigned under the U.S. Standard Industrial Classi-

Table 3.1   Company-financed research and development as a percentage of sales in the parent industries of the case study products, 1977

Case study products	R&D as % of sales
Wet shaving apparatus (part of SIC 3421)	3.3
Television sets etc. (part of SIC 3651)	1.6
Automobile and truck tires (SIC 3011)	2.0
Electronic calculators (part of SIC 3574)	7.3
Commercial airliners (part of SIC 3721[a])	3.3
Cameras and film (part of SIC 3861)	7.7
Diagnostic imaging equipment (part of SIC 3693)	3.7
Digital switches (part of SIC 3661)	4.9
Optical fibers for cables (part of SIC 3229)	4.2
Facsimile machines (part of SIC 3661)	4.9
Earth-moving machinery (part of SIC 3531)	3.1
All manufacturing industries	
Mean[b]	1.2
Median[b]	0.7

Source: U.S. Federal Trade Commission, Bureau of Economics, *Statistical Report: Annual Line of Business Report, 1977* (Washington, D.C., April 1985), Table 2-7. SIC codes are from the 1972 Standard Industrial Classification.

a. SIC 3721 includes a preponderance of military aircraft.

b. Calculated for 449 four-digit manufacturing industries with data derived from Federal Trade Commission reports for 221 more broadly defined industry groups.

fication had higher ratios of company-financed research and development to sales than the average for all manufacturing industries (see Table 3.1). Thus, the subject industries had a relatively high, but widely varying, propensity to develop and implement new technology. Nearly all experienced above-average growth of imports between 1963 and 1986 (see Table 3.2). The mean growth in imports as a percent of domestic output for the eleven parent industries was 1.75 percentage points per year; the median was 0.70 points, compared with 0.44 points for the U.S. manufacturing sector as a whole.[1] (In some cases, we shall see, import trends in the specific product lines differed appreciably from those in their broader four-digit parent industries, but the Table 3.2 data provide a useful and historically comparable first approximation.) The eleven parent industries' mean *level* of imports as a percentage of domestic output in 1981–

---

1. These estimated annual changes are derived from ordinary least squares regressions on a linear time trend.

Table 3.2   Imports as a percentage of domestic output value and annual import growth for the parent industries of the case study products, 1963–1986

Case study products	Imports as % of output				Annual growth[a]
	1963–86	1963–70	1971–80	1981–86	
Wet shaving apparatus	14.7	9.4	15.1	21.2	0.70
Television sets etc.	64.1	19.9	62.8	125.3	5.96
Automobile and truck tires	10.0	2.8	11.6	17.1	0.87
Electronic calculators	87.6	39.1	69.5	182.4	8.00
Commercial airliners	2.4	0.9	2.0	4.9	0.22
Cameras and film	10.2	4.9	9.6	18.4	0.78
Diagnostic imaging equipment	15.1	10.7	16.4	18.7	0.48
Digital switches	4.1	1.3	2.6	10.2	0.48
Optical fibers for cables	10.3	4.7	9.0	20.0	0.85
Facsimile machines	4.1	1.3	2.6	20.0	0.85
Earth-moving machinery	4.7	1.8	3.9	10.0	0.47

a. In percentage points per year, derived from a regression of annual imports as a percentage of output on a time trend variable.

1986 was 39.7 percent and the median 18.4 percent, compared with 11.5 percent for all manufacturing. Most of the sample industries also had substantial export levels, suggesting the applicability of intra-industry trade theories (see Table 3.3).

The case studies were prepared under the author's guidance by Judy Chevalier,[2] an economics graduate student at the Massachusetts Institute of Technology; Andreas Beckmann[3] and Anne McCormick,[4] undergraduate students at Swarthmore College; and Richard Gross,[5] John Lonnquist,[6] and Michele Rivard,[7] graduate students at the John F. Kennedy School, Harvard University. Three case studies were supplemented by materials from term papers written by Swarthmore College and Harvard University students.[8] Matthew Barmack provided general research assistance. Invaluable in the case study research were Harvard Business

2. Wet shaving apparatus, television sets, medical diagnostic imaging equipment, and facsimile machines.
3. Tires, earth moving equipment, and digital switches.
4. Cameras and film and airliners.
5. Electronic calculators.
6. Optical fibers and cables.
7. Facsimile machines.
8. John Haegele (April 1988) and Rishi Reddi (May 1987) wrote VCR studies; Jim Brumby (May 1991), an analysis of the Gillette Sensor razor development; and Ranjit Lamech (May 1991), a study of facsimile device development.

Table 3.3   Exports as a percentage of domestic output value and annual export growth
for the parent industries of the case study products, 1963–1986

	Exports as % of output				Annual growth[a]
Case study products	1963–86	1963–70	1971–80	1981–86	
Wet shaving apparatus	4.4	3.6	4.8	4.8	0.10
Television sets etc.	8.7	3.8	10.8	11.8	0.51
Automobile and truck tires	3.5	2.7	3.8	4.0	0.08
Electronic calculators	25.9	25.2	28.5	22.7	−0.04
Commercial airliners	35.3	13.3	36.0	63.6	2.62
Cameras and film	13.1	9.7	14.9	14.7	0.33
Diagnostic imaging equipment	24.2	18.2	26.8	27.8	0.67
Digital switches	3.6	2.0	3.6	5.8	0.22
Optical fibers for cables	7.0	6.3	7.9	6.6	0.05
Facsimile machines	3.6	2.0	3.6	5.8	0.22
Earth-moving machinery	34.0	31.1	35.8	34.9	0.21

a. In percentage points per year, derived as explained in Table 3.2.

School case studies and other published histories, company annual reports, and the extensive array of materials cited in the domestic and
international versions of the *Funk & Scott Index,* which categorizes trade
journal articles by company and Standard Industrial Classification code.
No field interviews were attempted. In four cases, however, comments
on draft versions were received from company executives involved in the
events studied. In all but two instances, the case studies were written up
as narrative histories much longer than the analytic summaries presented
here. Access to copies of the original case studies can be provided on
request.

The Individual Product Histories

Wet Shaving Apparatus

Building upon King C. Gillette's 1903 invention of the safety razor, which
gradually displaced the straight razor, the Gillette Company grew to become the dominant manufacturer of razors and blades in much of the
non-Communist world.[9] Gillette pursued a classic "tying" strategy: it

9. A significant exception is Japan, where U.S.-based Schick had a 70 percent market
share, leading Gillette and a Japanese company.

sold the razors themselves at very low prices, but earned generous profits from a continuing stream of blade sales. Even though others offered blades that fitted the Gillette holders, Gillette's share of the U.S. blade market remained at approximately 70 percent in the early 1960s. It was then that the first important foreign challenge to Gillette's position came.

The possibilities of stainless steel blades were widely recognized, and the major blade manufacturers were conducting desultory research on the concept. Gillette had introduced "Kroman" stainless steel blades in 1928, but quality problems led to the product's withdrawal in 1932. In the ensuing decades, each blade maker was reluctant to introduce stainless blades because they were expected to provide at least three times as many shaves as their carbon steel counterparts, and it was doubted that they could be priced in such a way as to yield profits as large as those derived from selling a much larger volume of carbon steel blades.[10] The first to upset this stalemate was England's Wilkinson Sword, which began exporting double-edge stainless steel blades to the United States in 1962, selling them at a retail price 2.2 times that of the shorter-lived Gillette Super Blue blade. Two U.S.-based competitors of Gillette soon followed suit. Eversharp-Schick test-marketed its Krona blade in February of 1963, with national introduction in August; and American Safety Razor introduced its stainless Personna and Pal blades in the spring of 1963. Gillette's first stainless blades came fourth chronologically, reaching the market in the fall of 1963.

Nevertheless, Gillette was not seriously discomfited by its delay. Wilkinson's penetration into the market was checked by a curious choice of marketing strategy. At first, its new blades were sold only by hardware stores and garden shops in conjunction with Wilkinson's line of garden tools, whose sale it hoped to promote through the attention-gathering power of stainless blades. Blade production capacity constraints also limited its ability to supply the U.S. market. By the time Wilkinson changed policies and distributed its blades through more conventional drug and grocery store channels, Gillette had struck back. Gillette's domestic rivals posed a more serious potential threat. In their rush to the market, however, they failed to solve all the quality control problems attending the adoption of a substantially new production technology. American

---

10. According to a contemporary account, "probably never in marketing history have companies entered a new product situation so reluctantly." "Cheek by Trowel," *Business Week,* December 22, 1962, p. 81.

Safety Razor (ASR) in particular was plagued by high rejection rates in production, and of the blades that reached the market, so many were defective that repeat purchases following a massive wave of first trials were disappointing.[11] ASR realized later that it had to redesign its product and retool its factory. Gillette, apparently confident that its marketing prowess would permit it to recoup, took time in late 1962 and much of 1963 to perfect the details of its product and production processes.[12] By 1965, when approximately four-fifths of all wet shavers had switched to stainless blades, and when Gillette launched a "Super Stainless" second generation, Gillette's share of U.S. stainless blade sales exceeded 50 percent. Wilkinson remained in the U.S. market, expanding its line of razor offerings, but its market share declined from 16 percent of razor blade sales in 1967 to less than 3 percent in 1989. Its blades were produced in England but packaged in the United States.

Although there were other innovations, most notably, the twin-blade razor, led by Gillette's Trac II in 1971, the next major challenge from abroad was the disposable razor. In 1975, Société Bic S.A. of France introduced a cheap throwaway single-blade razor in Europe. Gillette responded quickly to protect both its European and its U.S. franchises.[13] In 1976, it commenced worldwide sales of two disposable razors, the Good News for men and the Daisy for women. In the United States, Gillette's disposable product launch preceded that of Bic later in 1976. However, Gillette chose a slightly different product strategy. Bic sold, and continues to sell, only a single-blade disposable razor, while Gillette focused exclusively on twin-blade products, which offered superior shaving quality but commanded retail prices roughly 50 percent more than those of the Bic disposable. Again, Gillette's calculated risk appears to have paid off. Its Good News razor was the top-selling disposable razor

---

11. See David J. Ravenscraft and F. M. Scherer, *Mergers, Sell-offs, and Economic Efficiency* (Washington, D.C.: Brookings, 1987), pp. 239–240.

12. See "Close Shave?" *Forbes*, February 1, 1963, pp. 14–15.

13. As one of the first multinational corporations, Gillette had factories in Germany, France, England, and Canada as well as in the United States by 1910. In 1990 most of its U.S. demand was filled from U.S. factories, although some double-edged blades were imported from a plant in France. Bic produced razors for the U.S. market at a plant in Connecticut.

The import and export share data in Tables 3.2 and 3.3 are influenced disproportionately by movements in cutlery and scissors trade, also included in SIC 3421. Between 1977 and 1982, razor and blade imports as a percent of domestic consumption remained within a 3–5 percent range, while exports varied from 4 to 9 percent of output.

in the United States in every year between 1976 and 1989.[14] Its total share of the U.S. disposable razor market in 1986 was approximately 60 percent, with Bic following in second place at roughly 24 percent by dollar volume (but, given its lower prices, more by unit volume).

Even though Gillette maintained its relative market position against the disposable razor threat, it was less than fully satisfied with the outcome. The pricing of disposable razors is sufficiently competitive that disposables yield their makers significantly lower profits per shave. It was estimated that in the late 1980s, Gillette realized gross profit margins of 8 to 10 cents from each sale of a disposable razor, but 25 to 30 cents from comparable Atra or Trac II blade sales.[15] Because disposables accounted for 20 percent of U.S. blade sales in 1978 and nearly 50 percent in 1988, Gillette's defense of its market position, however successful, in effect cannibalized the profits that might have been gained from conventional blade sales, absent disposable razor competition.[16]

Gillette's response was to spend an unprecedented $75 million on research and development (as well as $125 million for production tooling and $110 million for roll-out advertising) on its Sensor razor system, introduced in January 1990. The Sensor mounts each of its twin blades on responsive springs to increase shaving closeness and comfort. The basic idea and design were conceived in Gillette's U.K. laboratories; production engineering was carried out in the United States; and initial production was located in Boston and in Berlin, Germany. Pre-production R&D required, inter alia, finding an economical way to join the blades to their holder. The solution, achieved after much trial and error, used lasers to weld the tiny assembly at 26 individual points. Following the strategy it had pursued successfully for the better part of a century, Gillette offered the Sensor razor at an extremely low suggested retail price of $3.75, anticipating that consumers' satisfaction with the product would ensure a continuing stream of blade purchases, whose premium prices would be protected by a dense network of patents both in the United States and abroad. Because the majority of Sensor buyers were already Gillette customers, it was unclear whether Gillette would

14. Gillette, *Annual Report: 1989*, p. 4.

15. "How a $4 Razor Ends Up Costing $300 Million," *Business Week,* January 29, 1990, p. 62.

16. See "Knocking Copy over Razors," *Boston Globe,* May 25, 1986, p. A-11; and "Can This 'Super Shaver' Save the Razor Business for Gillette?" *Adweek's Marketing Week,* August 14, 1989, p. 2.

gain much market share in conventional wet shaving devices through its innovation.[17] Gillette's aspiration, however, was that Sensor's superior properties would win consumers away from disposable razor use. Early reports revealed that 29 percent of Sensor buyers were previously disposable razor users—half switching from Gillette products and the other half from those of competitors such as Bic.[18] Despite the launch of a superficially similar but inferior product by Schick in November 1990, Gillette's profits rose appreciably in late 1990 and early 1991, mainly on the strength of Sensor sales.

Television and VCR Sets

Although many others participated, the Radio Corporation of America (RCA) played a central role in the development of the U.S. television manufacturing industry and in its later near capitulation to foreign competition. RCA was (with Dumont) a pioneer in developing black-and-white television during the 1930s. Working in tandem with its NBC broadcasting affiliate, it spurred the development of the medium as civilian activity resumed following World War II. In 1947, 80 percent of the 179,000 television sets sold in the United States were manufactured by RCA. In that year, however, both RCA and a competing network, CBS, had developed systems for the next logical step in television—color broadcasting. The RCA color system was designed to broadcast signals that could be viewed by existing black-and-white sets (to be sure, in black and white only) as well as by new color sets, whereas the CBS system was not black-and-white compatible. Both proposed systems were rejected by the Federal Communications Commission (FCC) in 1947 on the ground that, at least for the time being, monochrome broadcasting remained technically superior. In that same year, RCA made a crucial decision to license any and all companies to use its patented black-and-white television technology and designs in making their own sets. Its reasoning was apparently that widespread diffusion of the technology would lead to more rapid saturation of the U.S. market with black-and-white sets, which in turn would force the FCC to reject CBS's incompatible color standard and accept

---

17. See "Gillette's New Razor Is Sharp But Stock Barely Budges on Worry Success May Fizzle," *Wall Street Journal,* February 21, 1990, p. C-2.

18. "Gillette's Sensor Has Revived the Personal Care Product Giant," *Boston Globe,* January 27, 1991, p. 83.

RCA's.[19] The immediate consequence was a surge of television set manufacturing, much new entry, and a drop in RCA's production market share to 30 percent in 1948 and 12 percent in 1950.[20] By the end of 1949, some 4.25 million television sets had been sold.

The Federal Communications Commission returned to the color question in 1949 and approved the CBS design, with broadcasting scheduled to commence in 1950. In response, RCA mounted a delaying action in the federal courts, organized black-and-white set manufacturers and other broadcasting networks in opposition to the new standard, and worked to improve its own system design, introducing a substantially improved version in 1951. CBS meanwhile experienced technical difficulties with its design and sold disappointingly few color sets. Noting RCA's technical improvements and the fact that 22 million monochrome receivers were in use by the end of 1952, the FCC in 1953 reversed its prior decision and approved the compatible RCA color standard, known as the NTSC standard (after the FCC's National Television Standards Committee that made the recommendation). The Japanese government adopted the same standard in 1960. In Europe, a protracted battle led to the adoption in 1966 of two different color standards, SECAM from France and PAL from West Germany, each incompatible with the other and with the U.S. NTSC standard. These differences served to retard international trade in color television apparatus.

Because the new color television sets were expensive, color programming was limited, and many homes in the United States already had serviceable black-and-white sets, color set sales grew slowly. Only in 1963, when 50 million homes had black-and-white sets, was the millionth color set installed. Even though RCA was required under a 1958 antitrust consent decree to license at modest royalty rates all its color television patents, most black-and-white set manufacturers were slow to embrace the new technology. In 1960, only RCA, Packard Bell, and Magnavox were producing color sets. As NBC in particular aggressively expanded its color programming, sales of color sets accelerated during the mid-1960s, spurring black-and-white manufacturers to switch. By 1965, RCA's share of the color set market had fallen to 40 percent, while late

---

19. See Robert Sobel, *RCA* (New York: Stein and Day, 1986), especially pp. 150–151; and "The Five-Year Color War," *Television and Radio Age,* September 28, 1987, pp. A66–A67.

20. Sobel, *RCA,* p. 151.

entrants Zenith and Motorola captured second and third place with 20 and 10 percent respectively. Because RCA's color technology was available through licensing, the newcomers were required to make only modest investments in research and development. Zenith, for example, spent just $30 million on its color entry, mostly for plant expansion. In contrast, RCA's investment as of 1962 was said to be $130 million.[21]

Japanese electronics firms obtained licenses to RCA's black-and-white technology in 1952 and 1953 and to its color television patents in 1960. Their sales in the U.S. market were at first slight, while efforts of U.S. firms to penetrate the Japanese markets were quelled by the refusal of the Ministry of International Trade and Industry (MITI) to grant foreign currency allocations and by Japanese manufacturers' success in persuading retailers not to distribute the U.S. products. The first significant Japanese inroads into the U.S. market came in 1962, when Sony (which in the 1950s had led the capture of the U.S. transistorized radio market) introduced a line of portable transistorized portable black-and-white "microtelevisions," with screen diagonals ranging from four to nine inches. RCA had previously introduced a seven-inch screen model and found it virtually unsalable. But as increasingly affluent consumers traded up to color and U.S. producers emphasized the production of more profitable color sets at the expense of black-and-white technology, a large and essentially unserved second TV set market niche opened up for the Japanese micros. By 1969, Sony alone had sold a million microtelevisions in the United States. Another key event was the effort in 1963 by RCA and Zenith to persuade Sears Roebuck to sell the television sets they supplied Sears under their own brand names rather than under the Sears name. Instead, Sears accepted an offer from Sanyo of Japan to supply monochrome sets at prices lower than those any American firm would quote. Between 1963 and 1977, Sears purchased 6.5 million sets from Sanyo and Toshiba. Other retailers followed suit, so that by 1976, 80 percent of all U.S. private-label television sales came from Japanese companies.

Although U.S. manufacturers were the first to design both monochrome and color televisions with transistors replacing slower-starting, shorter-lived electron tubes, they failed to implement the newer technology in their main product offerings. By 1963, most Japanese sets exported to the United States were transistorized, whereas U.S. producers did not move to complete transistorization of their monochrome sets until the

---

21. Sobel, *RCA*, pp. 163 and 166.

late 1960s. Motorola introduced the first all-solid-state color television set in 1968, but a substantial price premium discouraged sales. The Japanese, meanwhile, moved aggressively to solid-state color designs, first in their home market and then for export sales. Through learning by doing, they reduced the cost and improved the quality of transistorized sets, gaining an advantage that permitted them to lead the move to solid-state color sets in the United States during the early 1970s. U.S. firms again lagged, with most of their transition to all-solid-state occurring during the mid-1970s.

Thus, following a late start, Japanese electronics firms had by the early 1970s moved to technological equality with their American counterparts on most relevant dimensions and had surpassed them in transistorization and reliability. This they achieved in part by investing in product research and development—and especially production process R&D—more aggressively than the U.S. firms. There appear to be four interacting reasons for the U.S. industry's failure to maintain its technological lead. First, the industry's movement into monochrome and then color television was greatly facilitated by easy access to RCA's technology, and as a result, other industry members were not compelled to build strong innovative capabilities of their own. Second, until the mid-1970s, U.S. firms systematically underestimated the seriousness of the Japanese challenge, in part because they had chosen to concentrate on making color sets, which was more lucrative, and had downplayed black-and-white technology, on which the Japanese had focused their export efforts during the 1960s. Third, by the early 1970s, the leading U.S. manufacturers were beginning to view color TV manufacturing as a mature business that did not warrant substantial, continuing investments in new technology.[22] Fourth, by 1976, when the Japanese share of U.S. color television set sales jumped to 36 percent from 18 percent the year before, the Japanese had achieved such a strong technological position, and the profitability of television set manufacturing in America had deteriorated so much, that it was difficult for the U.S. producers to finance a come-from-behind effort.[23]

---

22. See "Zenith and the Color Television Fight," Harvard Business School case study 383-070 (1982), p. 4.

23. Out of some 234 manufacturing industries for which FTC Line of Business survey data are available, radio and television set manufacturing ranked 209th in return on assets in 1975, 193rd in 1976, and 203rd in 1977. The industry's R&D/sales ratio fell from 2.0 percent in 1975 to 1.7 percent in 1976 and 1.6 percent in 1977.

The question remains, why didn't the U.S. industry's traditional technical leader, RCA, sustain a technological effort sufficient to force the pace for its domestic rivals and perhaps, assuming a continuation of past licensing policies, to help them stay abreast? The answer appears to be that RCA's attention and resources were diverted elsewhere, to a massive and ultimately unsuccessful campaign to become an important computer supplier. An RCA executive later admitted, "We shot a whole generation of research and engineering on computers and starved the real cash cow—color TV—to do it."[24] In addition to its large losses in computers, RCA lost appreciable sums in its efforts to develop consumer video recording technology—a story to which we return later. RCA also suffered less from increasing import inroads than did its U.S. rivals, because most Japanese firms continued to pay royalties on U.S. sales of sets licensed under RCA's color television patent portfolio.[25]

The position of U.S. television set makers was aggravated by the aggressive pricing tactics used by Japanese firms to seize the American market. Beginning in the 1950s, with the encouragement of MITI, Japanese consumer electronics makers formed an export cartel whose function, at least for television apparatus, was to keep prices in the domestic market high while encouraging vigorous exports at lower prices, with export "check price" floors set to minimize the likelihood of successful anti-dumping actions by target nations.[26] This was, of course, a classic "dumping" price discrimination strategy. The Japanese producers evidently "chiseled" on their agreements by undercutting the check prices through secret rebates to U.S. distributors. Although the evidence from various law suits is confused, it would appear that the prices of exports to the United States did not fall below marginal costs, and the leading producer, Sony, earned substantial profits in most years. U.S. television manufacturers initiated a series of anti-dumping and antitrust actions against the Japanese firms, but the first cases foundered as a result of weaknesses in the pre-1974 U.S. trade law, and later actions bogged

---

24. Peter Nulty, "A Peacemaker Comes to RCA," *Fortune,* May 4, 1981, p. 140. See also Sobel, *RCA,* p. 210: "RCA officials conceded that quality control in television declined in the late 1960s and that the corporation had delayed the introduction of transistorized models too long so as to squeeze all it could from tube-powered sales."

25. See "Zenith and the Color Television Fight," p. 5, which estimates RCA's total 1977 royalty earnings at $70 million.

26. See Clyde Prestowitz, *Trading Places: How We Allowed Japan to Take the Lead* (New York: Basic, 1988), pp. 199–206; F. M. Scherer and David Ross, *Industrial Market Structure and Economic Performance* (Boston: Houghton-Mifflin, 1990), pp. 469–472.

down in procedural complexity and were ultimately rebuffed by a sharply divided U.S. Supreme Court.[27]

Meanwhile, pressures from domestic television makers led to the negotiation in 1977 of an orderly marketing agreement with Japan, which cut Japanese exports of complete television sets to the United States to 1.56 million units per year (plus 190,000 unassembled units) from 1977 through 1980—a reduction from the 2.5 million sets exported in 1976. This induced three further reactions. First, Japanese companies shifted some of their manufacturing activities to the United States—Sony, for example, by expanding a California plant opened in 1972 and Matsushita (Panasonic-Quasar) by expanding the television production facilities it acquired from Motorola in 1974. Second, the Japanese greatly increased their export of television set parts from virtually zero in 1976 to an estimated three million equivalent units per year. Final assembly in the United States followed. Third, the limits on exports from Japan triggered a substantial increase in exports to the United States from Korea and Taiwan, especially for the private-label outlets through which the Japanese had gained their initial foothold. The orderly marketing agreement with Japan was thereupon extended to those two nations, continuing through June 1982.

Although the dollar's falling value during the late 1980s increased the profitability of television assembly in the United States, much of the work was left to be done by companies with a home base overseas. Magnavox was acquired in 1974 by Philips of the Netherlands, whose 1980 share of European Common Market television production was 30 percent. In 1980, Philips purchased the Sylvania and Philco brands. In 1987 Thomson of France, with a 10 percent share of European Community color television sales, purchased the consumer electronics division of General Electric, which had absorbed RCA, including its television operations, a year earlier. In 1990, only Zenith and Emerson remained as significant U.S.-owned suppliers to the American market.

The history of the television set's companion, the video recorder, can be analyzed more briefly.[28] Ampex of the United States demonstrated the

---

27. *Matsushita Electric Industrial Co. Ltd. et al.* v. *Zenith Radio Corp. et al.,* 475 U.S. 574 (1986). See also "Zenith and the Color Television Fight," pp. 6–13; and David Schwartzman, "The Japanese Television Conspiracy," manuscript, New School for Social Research, 1991.

28. The definitive histories are Margaret B. W. Graham, *RCA and the VideoDisc: The Business of Research* (Cambridge: Cambridge University Press, 1986); and Richard S.

first successful video signal tape recorder in 1956, but it and its successors were complex devices useful mainly to broadcasting studios. A potential demand for portable devices usable in educational, marketing, and even home environments was recognized at an early date. In 1960, Ampex and Sony jointly developed a transistorized portable version of the Ampex studio recorder. It, however, and similar machines devised by CBS, RCA, and the newcomer Cartridge Television, Inc. (bankrupted in 1973) were too bulky and expensive for consumer use, unreliable, otherwise ill suited to the mass market, or held back by fear that giving consumers the ability to record as well as play back video signals would engender copyright infringement problems. In 1970 Ampex began a two-year joint venture with Toshiba of Japan on a cartridge recording and playing machine, but the partners were unable to solve the problems of manufacturing their design sufficiently well to commence mass production. After absorbing a $90 million loss, Ampex chose to abandon the price-sensitive consumer market and concentrate on studio equipment. Thus, by the mid-1970s, RCA remained the sole U.S. firm with strong interest in video player-recorders for consumers.[29] Its interest was mirrored by Sony and others in Japan and by Philips in Holland.

During the late 1960s, RCA had been proceeding along multiple technical paths toward a consumer-oriented video player. It considered magnetic tapes, laser-decoded holographic tapes, and three different disc techniques. A holographic system was demonstrated with great fanfare in 1969, but by the early 1970s, the search had focused on magnetic tape versus discs read through capacitance variations by a stylus. Losses on RCA's computer operations forced a choice in 1974. It embodied important technical and marketing assumptions. On the technical side, RCA engineers believed that the pickup head on a mass-market tape recorder-player had to be held to a tolerance of two-tenths of a mil (that is, two ten-thousandths of an inch), but the best available manufacturing technology could attain tolerances of only a half mil.[30] Disc-based technology

Rosenbloom and Michael A. Cusumano, "Technological Pioneering: The Birth of the VCR Industry," *California Management Review*, 29 (Summer 1987), pp. 51–76. See also "The World VCR Industry," Harvard Business School case study 9-387-098 (1987).

29. Although it did not produce consumer-oriented video recorders, Ampex received substantial royalties from patents covering key concepts embodied in the components of VCRs sold by Japanese companies.

30. "The Anatomy of RCA's VideoDisc Failure," *Business Week*, April 23, 1984, pp. 89–90.

appeared more feasible. For this and other reasons, disc systems were also expected to be much less expensive, entering the consumer market at prices below $500, compared with the minimum $995 price anticipated for a tape-based system.[31] The disc approach had the disadvantage of permitting only playback and not recording by consumers, but RCA executives believed (in hindsight, correctly) that there would be a strong consumer market for means of viewing at will factory-recorded video programs, so they were disposed to sacrifice the self-recording attribute. This view was reinforced by studies showing that at high production volumes, video disc records could be produced at significantly lower unit cost than could tape recordings. In their dual role as broadcasters of copyrighted television materials as well as manufacturers, RCA executives may also have been influenced by a desire to avoid the legal battles that video recording eventually triggered.[32] Thus, RCA chose to focus its resources on developing a video disc system.

The Japanese and Dutch contenders decided differently. In part cooperatively but mostly independently, Sony and JVC, a subsidiary of Matsushita, painstakingly tackled the engineering problems blocking the path to a portable video cassette recorder. Between 1965 and 1974, both firms had made repeated attempts to develop and market recorders, but their designs were too limited in capability and too costly to tap the mass consumer market successfully.[33] They refused to be discouraged by failure, however. Sony elected in the early 1970s to build multiple prototypes embodying ten different design concepts before settling on a preferred solution. The problem of encoding sufficient information on a narrow tape was solved by recording diagonally rather than perpendicular to the tape's travel, as had been accepted past practice. As the effort moved forward, Sony's engineers whittled away the gap between current best practice and the required 0.2 mil pickup head tolerance at a rate averaging one-hundredth of a mil per month.[34] In December of 1974, a dispute erupted when Sony invited JVC to license and produce Sony's completed Betamax design on terms JVC considered unacceptable. Moreover, JVC

---

31. Graham, *RCA and the VideoDisc,* p. 150.

32. *Universal* v. *Sony et al.,* 659 F. 2d 963 (1981), 465 U.S. 112 (1984). See also James Lardner, *Fast Forward: Hollywood, the Japanese, and the VCR Wars* (New York: Norton, 1987).

33. Sony did successfully sell tape players for airliner motion picture presentation beginning in 1962.

34. "The Anatomy of RCA's VideoDisc Failure," pp. 89–90.

executives thought Sony's maximum one-hour playing time was too short. JVC therefore chose to continue its own independent development. As a result, Sony introduced its Betamax to the Japanese market in April 1975 and then entered the U.S. market in 1976. Joined by six Japanese manufacturer-licensees, JVC followed with its VHS system, providing an initial two-hour recording capability, in September 1976 for Japan and in January 1977 in the United States. Philips, which had been marketing a cruder VCR in Europe as early as 1972, reacted to the Japanese developments by improving its design and introducing its V-2000 machine in 1979—a year after the first marketing of Japanese recorders in Europe. Since each system used different tape formats, a battle over standards ensued, with JVC's VHS system emerging as the ultimate winner.[35]

Meanwhile, RCA dithered. By March 1975, it had advanced sufficiently far in the development of its SelectaVision videodisc system that it demonstrated a working prototype successfully to the press. As the technology stood during the next two years, the RCA system could be produced and sold at considerably lower prices than those commanded by VCR systems, whose list prices were in the $1,000–$1,300 range, and the reproduction quality of its images was considered superior. But RCA was experiencing problems converting its concepts to a design suitable for full-scale production, and it feared reliability problems with its pickup stylus–disc interface. With a new chief executive not educated in science or engineering at its helm, RCA cautiously held back commencing full-scale video disc production and marketing until March 1981, hedging its bet by marketing under its own brand name VHS tape-based machines produced to its specifications by Japanese firms. As the Japanese firms expanded the production of their VHS and Betamax systems and improved their technical capabilities, substantial cost reductions were achieved, facilitating a 50 percent reduction in VCR prices by the early 1980s and even sharper declines when Korean and Taiwanese clones reached the market in 1986. These learning-by-doing economies and competition-induced price reductions eliminated the cost advantage the poorly received RCA videodisc system had originally anticipated, and as a result, RCA abandoned its SelectaVision line in 1984, writing off losses of $580 million.[36]

---

35. See Peter Grindley, "Product Standards and Market Development: The Case of Video Cassette Recorders," London Business School Centre for Business Strategy working paper, May 1989.

36. "Splitting Up RCA," *Fortune,* March 22, 1982, p. 62; and "The Anatomy of RCA's VideoDisc Failure," pp. 89–90.

In its work on SelectaVision disc technology, RCA investigated both laser sensing techniques and a capacitive stylus pickup cartridge, favoring the latter.[37] In a sense, this too turned out to be backing the wrong horse. Laser disc techniques were the basis of the next important step in consumer electronic products, the compact disc audio player, introduced in 1981. The first CD players came from a joint development effort between Sony of Japan and the American subsidiary of Holland's Philips.[38] Germany's AEG-Telefunken had a different prototype under development, but the Sony-Philips approach became accepted as the industry standard, and so other firms have operated under Sony-Philips licenses.

Automobile Tires

The most important innovation of the past half century in the placid tire industry is the radial tire. It was invented by Dunlop of Great Britain but introduced commercially by Michelin of France in 1948. Steel-belted radial tires offered more than twice the average running life of the previous standard, bias-ply tires, along with greater safety, more sensitive road handling, and superior rolling efficiency (yielding fuel efficiency improvements of up to 10 percent). Radials swept the European market during the 1950s and 1960s, propelling Michelin from a position of modest sales, mostly in France, to the leading position in the European Common Market. Goodyear, at the time the world's largest tire maker, introduced its first radial tire to Europe in 1958, followed two years later by Uniroyal.

Michelin invaded the U.S. market, which had previously been dominated by a few domestic companies, in the mid-1960s. In 1966, it won an important account with Sears Roebuck. The first American manufacturer to offer radials on the U.S. market was the B. F. Goodrich Company, in 1965. Despite an aggressive marketing campaign, its sales were disappointing, and the other U.S. tire makers refrained from following suit. Meanwhile, Michelin gradually built its U.S. sales, first by importing from France and then in the 1960s from Canadian plants.

Radials were only slowly accepted in the United States because Americans were accustomed to the smoother ride that resulted from bias-ply tires' greater flexing propensity—an attribute that simultaneously ac-

---

37. See Michael L. Dertouzos, Richard K. Lester, and Robert M. Solow, *Made in America* (Cambridge: MIT Press, 1989), p. 227.

38. Philips had previously developed a laser disc video player system. It too was abandoned to emphasize tape systems.

counted for the bias tires' shorter life. U.S. automobile manufacturers were reluctant to redesign their suspensions for the different handling properties of radials. In addition, American drivers, facing much lower gasoline prices than their heavily taxed European counterparts, were insensitive to the superior mileage advantage of radials. And few American states allowed their citizens to race along the public highways at the 100-mph speeds common in France and Germany, so the safety merits of radials were also less appreciated.

The technological response of American tire makers to the radial's appearance was therefore a compromise—the bias-belted tire, whose handling, efficiency, and life were roughly midway between those of the bias-ply tire and the belted radial. Goodyear was first, introducing its wildly successful Polyglas line in 1967. Others imitated quickly, so that by 1972 bias-belted tires had captured more than half the U.S. market.

The radial, however, continued its slow advance. Auto manufacturers began placing greater stress on safety and sharp handling, and they specified radial tires for some of their top-of-the-line offerings. Uniroyal joined Goodrich as a domestic radial producer in 1971. Goodyear's 1971 annual report observed, "Steel-belted radials and the company's development of an all-steel radial are gaining attention in this relatively young U.S. market." In 1972, Goodyear introduced four lines of radials and supported the launch with a massive advertising campaign. The timing was propitious: with the first OPEC oil shock of 1973–74, consumers became more mileage-conscious, and radial tire sales accelerated. In 1973, radials won 18 percent of original equipment sales and 13 percent of replacement sales. By 1980, the shares had increased to 83 percent and 54 percent, respectively.

Spurred in part by a 6.6 percent countervailing duty levied in 1973 on imports from its Canadian plants, which the U.S. Treasury Department found to have received Canadian government subsidies, Michelin opened a large plant in South Carolina during 1975. Even though its major U.S. rivals now offered radials, Michelin's long experience gave it a quality advantage, and its reputation for superior quality allowed it to command premium prices while gaining market share. The U.S. firms took note. "What worries Akron," a business journal observed in 1976, "is Michelin's single-minded emphasis on quality and innovation."[39] In that same year, Goodyear reorganized its research and development section

---

39. "Michelin Goes American," *Business Week*, July 26, 1976, p. 56.

to emphasize product innovation. In 1979, it launched "Operation Perfection," discarding all sub-par tires rather than selling them in the replacement market. Other firms worked to improve their quality, although Firestone suffered severe losses when its "500" radial had to be recalled owing to widely publicized failures in use. By the late 1970s the quality gap had been narrowed, even if not eliminated, and Michelin found itself compelled to stress price in advertisements and bid for a large Ford contract at rock-bottom prices. As an industry executive asserted in 1980, "Michelin has been fat, dumb, and happy for a long time. They don't make the best radial in the world [any longer]."[40] The U.S. tire makers also intensified the pace of product innovation, introducing during the 1970s puncture-sealing linings, all-weather treads, lightweight spare tires, and improved materials.

Meeting Michelin's technological challenge did not end the industry's problems, however. By greatly lengthening tire lives, the shift to bias-belted and then belted radial tires reduced the number of replacement tires required by a given car fleet. Excess capacity emerged, precipitating price warfare and the closure of many old plants. U.S. auto makers were unprepared for the second energy shock of 1979–1981, and renewed demand for small cars was satisfied by a surge of foreign autos, especially from Japan, carrying foreign-made tires. Tire imports rose from 7.6 percent of domestic demand in 1975 to 11.2 percent in 1980 and 23.2 percent in 1985. The sharp recession of 1982 reduced domestic original equipment demand further. The tire makers took advantage of relaxed antitrust enforcement to consolidate their deteriorating positions. Bridgestone, the leading Japanese tire producer, established a U.S. beachhead by purchasing a Firestone radial truck tire plant in 1982, and in 1988 it acquired the balance of third-ranking Firestone. Bridgestone's rival Sumitomo Rubber acquired control of Britain's Dunlop Tire, including its U.S. operations, in 1986. In 1986 Uniroyal and Goodrich merged their operations, which were in turn purchased by Michelin in 1990, propelling Michelin to first place among the Western world's tire makers.[41] Continental Reifen AG

---

40. "Michelin: Spinning Its Wheels in the Competitive U.S. Market," *Business Week,* December 1, 1980, p. 119.

41. On the difficulties that followed, see "Michelin: The High Cost of Being a Big Wheel," *Business Week,* November 5, 1990, p. 66; and "The Tyre Industry's Costly Obsession with Size," *The Economist,* June 8, 1991, p. 65. Bridgestone also had teething problems. See "Now Akira Yeiri Really Has to Burn Rubber," *Business Week,* May 27, 1991, p. 72; and "Additional Layoffs Set for Bridgestone Units," *New York Times,* August 1, 1991, p. D4.

of Germany acquired General Tire in 1987, Pirelli of Italy acquired Armstrong Tire in 1988, and Yokohama Rubber bought Mohawk Rubber in 1989. By 1990 only two companies, Goodyear and Cooper, remained as major domestically owned participants in the U.S. market.[42]

Electronic Calculators

Up to the mid-1960s, non-routine but moderately complex arithmetic calculations were typically performed on mechanical calculators—typewriter-size boxes full of finely machined gears and ratchets that whirred and ground out a solution to the problem punched in by keyboard. In the United States, the dominant specialized calculator producers were Friden, Marchant, Monroe, and Victor. If nonlinear operations on trigonometric or exponential functions were required, one used a slide rule, or, to achieve higher accuracy, consulted published tables.

The first shot in the electronic calculator revolution was fired in 1962 by a U.S. aerospace components manufacturer, Wyle Laboratories.[43] Wyle's transistorized device had programming and memory capabilities substantially beyond those of mechanical calculators and was priced at $4,000, well above the $800 for which a standard Friden calculator could be obtained. The Wyle machine achieved only limited market acceptance, and Wyle withdrew from the calculator business in 1967. However, its achievement impressed the mechanical calculator makers, four of whom initiated their own development projects, leading to the introduction of new electronic desk-top calculators between 1964 and 1967. But Friden and its peers shared a common problem—they knew very little about electronic, as distinguished from electromechanical, technologies. Considering the new electronic machines a relatively unimportant adjunct to their mechanical mainstays, they decided not to invest in a wholly new type of manufacturing, but contracted with such Japanese electronic goods producers as Hitachi, Toshiba/Casio, Canon, and Systek to provide calculators for sale through U.S. office machine distribution channels.

---

42. See "Why Tiremakers Are Still Spinning Their Wheels," *Business Week,* February 26, 1990, pp. 62–63; and "After a Year of Spinning Its Wheels, Goodyear Gets a Retread," *Business Week,* March 26, 1990.

43. This paragraph and the next draw heavily upon Badiul A. Majumdar, "An Empirical Study of Diffusion of Technology: The Case of Electronic Calculators," undated manuscript.

In addition to experience producing electronic goods, the Japanese firms had another advantage. The early electronic calculators contained hundreds of individual transistors and other components, whose assembly was highly labor-intensive. Because at prevailing exchange rates Japanese wages were much lower than U.S. labor costs, it was significantly less expensive to assemble the new calculators in Japan. The Japanese firms quickly mastered the new calculator technology and designed their own models. The American mechanical calculator makers had assumed that their well-established sales and service networks would be sufficient to preserve their advantage against independent Japanese competition, but in this they erred. Mechanical calculators, with their many moving parts, broke down frequently and needed fast, reliable service, but transistorized devices, once "burned in," required little post-sale service, and so traditional distribution channels could be bypassed. Offering highly price-competitive calculators of advanced design, the Japanese firms overwhelmed the American incumbents, winning an estimated 79 to 85 percent of all U.S. electronic calculator unit sales by 1970.[44] Eventually all of the specialized U.S. mechanical calculator producers and also slide rule maker Keuffel and Esser, which attempted a belated entry into electronic devices, exited from the calculator business.

Rapid advances in semiconductor technology brought important changes to the newly emerging calculator industry. The advent of integrated circuits reduced the number of electronic components to a few dozen. In 1968, Hewlett-Packard, traditionally an instrument maker, entered the field with a miniaturized device requiring only four metal oxide-silicon integrated circuit chips. It soon became the leading U.S. source for electronic calculators. Casio of Japan took the next crucial step in August 1972, marketing the first true mini-calculator, built around a single MOS/LSI chip of U.S. design.[45]

As the number of discrete components assembled to make a calculator fell, so also did the importance of labor costs—to as little as 10 percent of total costs with the introduction of Casio's single-chip mini-calculator.[46]

---

44. The lower estimate is from Majumdar, "Empirical Study," p. 25, the higher estimate from Franco Malerba, *The Semiconductor Business* (Madison: University of Wisconsin Press, 1985), p. 207.

45. Philips of Netherlands also developed MOS/LSI chips for desk calculators during the late 1960s, but they were apparently not very successful. See Malerba, *Semiconductor Business,* p. 113.

46. See Badiul Majumdar, "Technology Transfers and International Competitiveness:

Integrated circuits, which Japanese assemblers had for the most part purchased from U.S. firms, rose commensurately as a cost component. Recognizing this, and fearful that their calculator-assembling customers might integrate backward into producing their own chips, American semiconductor manufacturers entered the calculator business. Texas Instruments introduced its first three calculator products in September 1972. The proximity of its product launch date to Casio's reflected the fact that both companies were working in parallel toward the same goal, rather than TI rapidly copying Casio. Chip-maker Rockwell International joined the fray with private-label products for Sears Roebuck in 1972 and added its own brands in 1974. National Semiconductor presented its Novus line in 1973. Hewlett-Packard and another U.S. mini-calculator leader, Bowmar, reacted by integrating backward into chip making. Japanese calculator assemblers also began turning to their Keiretsu affiliates for chips. But with their well-established leadership position in semiconductors, companies such as Texas Instruments believed that comparative advantage had shifted in their favor. Japan's share of U.S. calculator sales fell to 25 percent in 1974, and an industry analyst observed, "No one in the industry worries about the Japanese anymore."[47]

Along with its vertical integration strategy and technological breakthroughs, Texas Instruments contributed a business strategy innovation. Advised by the Boston Consulting Group, it openly avowed learning curve pricing—that is, setting prices below cost early in the product life cycle, with the goal of stimulating demand and winning the lion's share of sales, thereby racing down the learning curve and becoming the industry's low-cost and ultimately most profitable producer.[48] Within a year of its entry, TI became the world's largest calculator vendor. Consumers benefitted as the average price of mini-calculators fell from $100 in late 1972 to $35 in 1975. Most of the numerous U.S. companies that had entered the industry during the early 1970s sustained massive losses and dropped out before the decade ended. But some participants held on tenaciously. The Japanese, in particular, retaining a cost advantage in

---

The Case of Electronic Calculators," *Journal of International Business Studies,* 11 (Fall 1980), pp. 103–111.

47. "The Semiconductor Becomes a New Marketing Force," *Business Week,* August 24, 1974, p. 39.

48. See ibid. and "Selling Business a Theory of Economics," *Business Week,* September 8, 1973, pp. 85–90. On learning curves generally, see the discussion accompanying Figure 2.3 in Chapter 2.

assembly operations and tapping rapidly improving domestic semiconductor sources, could not be dislodged. As tough price competition continued, Texas Instruments found its margins squeezed severely. Its Fort Walton, Florida, calculator plant was closed, and in the second quarter of 1975 it reported calculator line write-offs of $16 million.[49] This setback precipitated a change of strategy. Following the pattern adopted earlier by Hewlett-Packard, TI successfully shifted its emphasis to top-of-the-line scientific/programmable calculators, where less aggressive niche-pricing strategies could be sustained. A concomitant was the virtual abandonment of the price-competitive simple calculator market to Japanese rivals, who have dominated it ever since.

Japanese electronic goods producers have displayed greater tenacity in learning curve pricing. Whether they learned the strategy from Texas Instruments or would have adopted it in any event is uncertain. What is clear is that they have pursued the strategy aggressively in other lines too—for instance, as we have seen, in color television sets during the mid-1970s and, much to the dismay of Texas Instruments and other U.S. chip makers, in DRAM semiconductors during the early 1980s.[50] A consequence has been the winning of substantial market shares in high-technology product lines once dominated by American companies.

Turbojet Airliners

The existence of systematic, well-behaved learning curves was first established conclusively using statistical data accumulated from World War II aircraft production.[51] On average, it was found, unit assembly cost fell by 20 percent with each doubling and redoubling of cumulative output. Commercial airliners are priced on an essentially level basis, abstracting

---

49. "Calculator Competition Helps Consumer, But Many Manufacturers Are Troubled," *Wall Street Journal,* January 14, 1975, p. 38; the survey of employment cutbacks in *Wall Street Journal,* January 23, 1975, p. 13; and "Texas Instruments' Earnings for 2nd Quarter Put under Microscope," *Wall Street Journal,* July 29, 1975, p. 37.

50. See Andrew R. Dick, "Learning by Doing and Dumping in the Semiconductor Industry," *Journal of Law and Economics,* 34 (April 1991), pp. 133–160; and Semiconductor Industry Association, *The International Microelectronic Challenge* (Cupertino, Calif., May 1981), pp. 2 and 19–20.

51. See Harold Asher, *Cost-Quantity Relationships in the Airframe Industry* (Santa Monica: RAND Corporation study R-291, July 1956); and Jack Hirshleifer, "The Firm's Cost Function: A Successful Reconstruction?" *Journal of Business,* 35 (July 1962), pp. 235–255.

from inflation and design changes, in the expectation that profits will begin emerging only when a sufficient volume has been accumulated to bring unit costs below unit revenues.

In the first decade following World War II, the supply of commercial airliners in the United States was dominated by Douglas Aircraft, with its propeller-driven DC-4, DC-6, and DC-7; and Lockheed, with its family of Constellations. The dawn of the jet age brought radical changes. The first commercial turbojet airliner was the Comet I, produced by DeHavilland of England and put into service in 1952. The design of its windows proved to be defective, and lethal explosive decompressions ended its sales at a total of 33 units. A redesigned version inaugurated in 1957 carried only 67 passengers and was overwhelmed by more commodious new competition. Cold war conditions left the Soviet Union's Tupolev 104, introduced in September 1956, with few free-world sales opportunities.

The first major success was the four-engine long-range Boeing 707, which entered commercial service in October 1957. Boeing began with the important advantage of vast experience gained in producing 1,373 B-47 bombers, using similar fuselage and wing-mounted engine pod designs,[52] as well as government contract support for a B-707 derivative, the KC-135 in-flight refueling tanker. When American Airlines ordered 30 aircraft after the 707 made its first flight in 1955, Douglas Aircraft reacted with an accelerated program to develop the DC-8, which entered service a year after the 707. Convair, another bomber manufacturer, followed with its 880 in May 1960. The DC-8 and 880 designs were very similar to that of the 707, and with little new to offer and a late start, they fell far short of the 969 B-707 units Boeing ultimately delivered. Among other things, Boeing's first-mover advantage stemmed from airlines' unwillingness to add new aircraft to their fleets, complicating pilot training and maintenance, unless the newcomer offered unique advantages over aircraft already purchased. Lockheed bet on an alternative technology with its turboprop Electra, which turned out to be the wrong choice and suffered also from fatal wing vibration failures.

Boeing consolidated its strong initial position with a series of additional models meeting other airline needs: the medium-haul tri-jet B-727, first

---

52. During the Korean war buildup, B-47 contracts were also awarded to Douglas Aircraft and Lockheed, but the quantities were much smaller, and Boeing insisted that those second sources be deprived of details on how its own production was organized.

delivered in 1964;[53] the much smaller medium-haul twin-jet B-737, inaugurated in 1968; and the spectacularly successful long-range jumbo B-747, appearing in 1969. These were modified and stretched in various ways to maintain Boeing as the Western world's leading commercial jet plane maker, commanding roughly two-thirds of that market's airliner sales during the 1960s and 1970s.

Following Boeing's initial successes, there were three main challenges to its dominance, one from the United States and two from abroad. (We ignore here short-hop commuter airliners, whose market in the United States was quite limited until airline deregulation occurred during the late 1970s, and which U.S. aircraft makers have for the most part ceded to European firms.) The domestic challenge came when both Douglas (acquired by McDonnell Aircraft) and Lockheed saw an unfilled need for a wide-body aircraft similar in principle to the B-747, but of less gargantuan dimensions and with superior aerodynamics. Each developed tail-mounted tri-jets, the Douglas DC-10, delivered first in August 1971, and the Lockheed 1011, which entered service in April 1972. But in offering essentially duplicative products, Douglas and Lockheed divided the market so that neither could achieve sufficient volume to cover its costs. As of 1985, Douglas had sold 379 DC-10s, Lockheed 248 L-1011s, and each lost considerable sums on the venture.[54]

Originally Douglas and Lockheed had intended to develop efficient medium-range twin-jets, but were persuaded by their customers to build in greater capabilities requiring a third engine. Still a gap in the market remained, especially after the first OPEC shock of 1973–74 raised jet fuel prices substantially and intensified the airlines' concern over energy efficiency.

Into this gap marched a new challenger, Airbus Industrie.[55] Stung by

---

53. The 727's tail-mounted engine configuration followed the twin-jet approach of the French Caravelle, which entered medium-haul service in 1959 and became the best-selling European airliner of the 1960s and 1970s, with 279 units delivered in total.

54. These and most other cumulative volume figures are drawn from a special industry survey, "Eternal Triangles," *The Economist,* June 1, 1985, p. 8. For a fuller analytic history, see John Newhouse, *The Sporty Game* (New York: Knopf, 1982). Historical experience suggests that airliner producers cannot expect to break even on their development, tooling, and high top-of-the-learning-curve costs until they have sold 400 to 600 units in a model series.

55. See Gernot Klepper, "Entry into the Market for Large Transport Aircraft," *European Economic Review,* 34 (June 1990), p. 777, for a diagram of how various airliner models were positioned on two-dimensional range and capacity coordinates.

the repeated failures of their national aerospace firms to sustain significant airliner sales in the world market, the governments of several European nations encouraged a collaborative effort to make Europe succeed
in what was considered a key high-technology field. Following four years
of inconclusive organizational and design effort, the resulting consortium,
Airbus Industrie, was formally established in December 1970. The principal participants were Aerospatiale S.A. of France, Messerschmitt-Böl-
kow-Blohm of West Germany, Fokker of Holland (which later dropped
to a subordinate role), and CASA of Spain. British Aerospace, which had
withdrawn in 1969 from the preliminary effort, rejoined in 1979. The
group's first product, the A-300, made its initial flight in October 1972 and
was commercially available in May 1974. Achieving superior efficiency
through aerodynamic innovations and two U.S.-designed, French-
assembled turbofan jet engines, it carried 250 passengers and had a range
of 1,550 miles, extended in later versions to 3,000 and then 3,500 miles.
Planning began on an advanced modification with lower passenger capacity (approximately 218 persons) but longer range, the A-310, in 1976. The
A-310 made its first flight in 1982 and entered service in 1983.

Boeing's first reaction to the wide-body tri-jet and Airbus competition
was to continue improving its B-727, 737, and 747 mainstays. However,
it began planning an advanced version, temporarily designated the 7X7,
and during the summer of 1976 it discussed collaboration on the project
with Airbus. The negotiations came to naught, as did discussions between
McDonnell-Douglas and Airbus for a collaborative effort that would have
restricted Airbus to the production of shorter-range aircraft. In July 1978
new and decisive steps were taken by both Airbus and Boeing. Airbus
received assurances of $1 billion in financing from its supporting govern-
ments, and so the A-310 program, already in the planning stages since
1976, was formally initiated. At the time, a total of 50 A-300s had been
delivered. Almost simultaneously, Boeing received a firm order for 30
new B-767 aircraft, a wide-body long-range twin-jet with a planned capacity of 216 passengers. Full-scale development was therefore commenced.
A few weeks later, firm orders were received for another new airliner, a
medium-range narrow-body twin jet carrying approximately 186 passengers, and so the B-757 program was launched. Boeing's company-
financed research and development expenditures soared from $222 mil-
lion in 1977 to a peak of $844 million in 1981 and then declined to $429
million in 1983 after the first commercial deliveries were made.

Meanwhile, McDonnell-Douglas struggled to keep its airliner offerings viable by augmenting its existing designs, leading in 1980 to the MD-80, a successor to its aging twin-engine DC-9, and in 1988 to the start of a $3.3 billion development and tooling effort on the MD-11, an improved version of the DC-10.[56] First deliveries of the MD-11 began in late 1990. In September 1990, McDonnell-Douglas announced preliminary plans to produce a new longer-range complement, the MD-12X. To finance the effort, it agreed in 1991 to sell a 40 percent interest in its commercial aircraft division to the Taiwan Aerospace Corporation.

March of 1984 brought one Airbus response to the initiatives of its U.S. rivals. Airbus announced its plan to build the twin-engine A-320, carrying 150 to 180 passengers 2,200 miles in a domestic version and 3,500 miles in an overseas model. Larger, longer-range extensions, the A-330 and four-engine A-340, were announced in June 1987. These new members of the family were designed to incorporate significant technological advances, including extensive use of lighter-weight fiber-resin structural materials in place of aluminum, "fly by wire" control mechanisms instead of the cables and hydraulic fluid lines used to control the movable wing and tail parts of earlier passenger airliners, and computerized controls designed to prevent maneuvers that exceeded the aircraft's aerodynamic limitations. The first A-320s began commercial service in 1988, when Airbus Industrie's share of Western world non-commuter airliner sales rose to 30 percent.

The technologies adopted in the A-320 and later models had been widely used in military aircraft developed during the 1970s. Boeing lagged Airbus in adopting them for three main reasons. First, Boeing's engineering philosophy was intrinsically conservative, tending to avoid new technologies unless they provided decisive advantages or were strongly desired by customers.[57] Second, the decline of oil (and jet fuel) prices during the mid-1980s from their record 1981 levels induced some relaxation of the airlines' demands for substantially increased fuel efficiency. Third and perhaps most important, with its huge experience producing conventionally designed airliners, Boeing had progressed much farther down relevant learning curves than any of its competitors. To derive the greatest benefit from that cost advantage, it attempted to maximize the

---

56. "McDonnell's New Jet Is Ready for Delivery," *New York Times*, November 9, 1990, p. D3; and "McDonnell Douglas: Disarmed and Dangerous," *The Economist*, January 19, 1991, p. 66.

57. Airline pilots had decidedly mixed emotions about surrendering their power over aircraft maneuvers to computers.

commonality of components among the members of its aircraft series and avoid changes that required it to jump to a wholly new learning curve, where it in effect had to begin at the top. With its newer models, Airbus operated on a new, more advantageous learning curve, but because of its relatively small cumulative volume, its position was well up that curve, with higher effective costs (depending also, from the user's perspective, on complementary jet fuel prices) than those of Boeing.[58]

Complicating the strategy choice was the fact that the success of Airbus was a high-priority objective of European Community nations, and to achieve their goals, they were willing to provide generous subsidies. How large those subsidies have been is a matter of debate. Airbus Industrie itself lost money in every year between 1970 and 1989.[59] It reported its first operating profit in 1990, and it was making loan repayments to European governments of approximately $10 million per delivered aircraft.[60] However, Airbus only assembles the parts transferred to it by its various national partners, and excessive transfer prices might conceivably cause a deficit for the assembler while yielding profits, concealed in annual reports covering multiple lines of business, to the partner firms. That the true losses and hence subsidies have not been overstated is suggested by the fact that the German government was anxious to limit its obligation to provide further subsidies in approving the merger of Messerschmitt-Bölkow-Blohm with Daimler-Benz in 1989. Thus the U.S. Department of Commerce 1989 estimate that the Airbus program had received cumulative subsidies totalling $16 billion appears plausible.[61] Jumping to the top of a new learning curve is less perilous when there is a safety net at hand to cover losses. European officials defended accusations of unfair subsidy by asserting that the U.S. airliner manufacturers also benefitted from military program subsidies, which was true for the 1950s, and that they expected the Airbus program eventually to become profitable as it moved down its own learning curves. They insisted also that the presence of

---

58. On Airbus learning curves, see Klepper, "Entry," p. 795.

59. See "Pressure Grows for U.S. to Penalize Airbus as Talks on European Subsidies Stall Again," *Wall Street Journal,* July 27, 1987, p. 5; and "Bailing Out Airbus," *Business Week,* December 18, 1989, p. 47.

60. In 1990, those repayments are said to have been approximately $500 million. "Airbus Payment," *New York Times,* June 12, 1991, p. D4.

61. "Bailing Out Airbus," p. 47. See also "Aid to Airbus Called Unfair in U.S. Study," *New York Times,* September 8, 1990, p. 17, which estimates out-of-pocket subsidies at $13.5 billion; "Is Airbus Pushing the Envelope?" *Business Week,* October 8, 1990, p. 42; and "Dissecting Airbus," *The Economist,* February 16, 1991, pp. 51–52.

Airbus injected important competition into markets that would otherwise be dominated by Boeing. Despite repeated attempts to resolve the dispute through conciliation,[62] the United States filed a formal complaint with GATT in March 1991 charging that government subsidies protected the West German partner from exchange rate fluctuation losses. Further actions were contemplated.

In any event, Boeing responded to the competitive pressure from Airbus (and less important, from McDonnell-Douglas) in 1990 by initiating development and sales efforts for a completely new twin-engine 350-seat aircraft, the B-777, with fly-by-wire technology, composite fiber structures, and other features embodied in the A-320 and its derivatives. Development and tooling costs were expected to total $4 billion.[63]

Government subsidy or lack thereof was decisive in determining where supersonic transports were produced. In 1966, Boeing won a U.S. government–sponsored competition to develop a supersonic transport. A fierce debate over the program's cost, economic viability, and environmental impact followed, and in 1971, governmental support was withdrawn. Boeing chose not to go forward on its own. Previously, the British and French governments had agreed in late 1962 to develop jointly a supersonic aircraft, appropriately named the Concorde. The development proceeded slowly, with the first prototype flight taking place in 1969 and the first unit entering commercial service in 1976. Substantial losses were incurred in development and production, operating costs were high, and concern over the aircraft's sonic boom prevented the Concorde from flying supersonically except over oceans and uninhabited areas. Because of these problems, only 14 of the 74 Concordes originally ordered were delivered for commercial use.

Cameras and Film

The Eastman Kodak Company has dominated U.S. and Western world markets for cameras and especially photographic films and print materials ever since 1880, when George Eastman built a large plant in Rochester to mass-produce his pioneering amateur camera and the films it required.

---

62. See "Airbus versus Boeing" (A) and (B), Harvard Business School case studies 9-386-193 (1986) and 9-388-145 (1988).

63. "How Boeing Does It," *Business Week,* July 9, 1990, pp. 46-50; and (on the first booked order) "It's Fat and Snazzy—and Worth Billions to Boeing," *Business Week,* October 29, 1990, p. 32.

By 1890, Kodak had become an early multinational enterprise, with production in England as well as in the United States. Between 1958 and 1968, Kodak's shares of U.S. "snapshot" camera, film, and color photographic paper sales averaged 68, 86, and 89 percent, respectively.[64] The company was nearly as successful overseas, commanding a 65 to 70 percent share of free-world color film sales.[65] Its failure to enter the Japanese market during the 1930s, however, left a void filled by the creation of Fuji Photo, which in the early 1980s held a 71 percent share of Japanese film sales, compared with only 13 percent for Kodak.[66]

Kodak's basic business strategy was similar in some respects to Gillette's, but with important differences in detail. The purchase of a camera in effect ties a consumer to purchasing a continuing stream of film, paper, and (except for buyers with their own darkrooms) photofinishing services. It was crucial therefore for Kodak to dominate the sale of film and paper, on which it realized profit margins estimated to exceed 50 percent.[67] This it did in several ways. It offered excellent quality. It tied the sale of color film to finishing services by providing the film and development service at a single "bundled" price until required to discontinue the practice and aid the formation of independent photofinishers in a 1954 antitrust decree. After that, it maintained excellent support for and rapport with independent film retailers and photofinishers. By refusing to supply film in formats other than those of its own choosing, it also impeded the entry of cameras using film formats that might weaken its own dominant position.[68]

In the sale of cameras, Kodak's record was more mixed. It excelled at supplying cameras aimed at a mass market interested mainly in simplicity and moderate quality.[69] During the 1940s and 1950s, its mainstays were the legendary Kodak Brownies. In 1963, it introduced the highly successful Instamatic 126, whose cartridge mechanism not only simplified loading but tied consumers more strongly to 35-mm film cartridges of Kodak

---

64. See James W. Brock, "Structural Monopoly, Technological Performance, and Predatory Innovation," *American Business Law Journal,* 21 (Fall 1983), p. 294.

65. See "Kodak's New Lean and Hungry Look," *Business Week,* May 30, 1983, p. 33.

66. "In Skies over Tokyo, Kodak and Fuji Fight Battle of the Blimps," *Wall Street Journal,* December 30, 1986, p. 1.

67. "Kodak Takes a Risky Leap into Consumer Video," *Business Week,* January 16, 1984, p. 92.

68. See Brock, "Structural Monopoly," pp. 299–300.

69. In this, Kodak apparently followed P. T. Barnum's little-known second law: "No one ever got rich by overestimating the taste of the American consumer."

origin. A much smaller Pocket Instamatic 110, using cartridges and 16-mm film of special format and design, enjoyed sales of more than 8 million units in the two years following its 1972 introduction. Ancillary improvements in easy-handling flashbulb technology helped fuel explosive growth in amateur photography, with Kodak as an important beneficiary.[70]

In the much smaller markets for professional and serious amateur cameras, Kodak was consistently a technological laggard. In the United States, Graflex led in supplying 4 × 5 inch and larger professional cameras until improvements in emulsion technology (contributed by Kodak) sharply reduced the demand for bulky, large-format instruments. Rolleiflex and Hasselblad provided the highest-quality 2¼ × 2¼ cameras, and in the 35-mm format, Leica of Germany took the lead in 1925, to be joined by Nikon and Canon of Japan during the 1950s. The Kodak 35 sold during the 1940s and 1950s was of markedly inferior quality compared to its German and Japanese competitors, and in 1970, Kodak discontinued the production of standard-format 35-mm cameras altogether. Kodak apparently chose not to compete aggressively in high-quality cameras because it was confident that its strong position in films and papers would allow it to sell most of the supplies for those cameras. And to the extent that expert photographers experimented with the film and paper of such European competitors as Agfa and Ansco, the high-quality market was in any case small, so the profits sacrificed by sharing it with others were at best modest.

Two courses of events, one on the camera front and the other in films, threatened the viability of Kodak's chosen strategy.[71] One began with the introduction of the Canon AE-1 camera in 1976. Its highly automated features made 35-mm picture-taking nearly as easy as with Kodak's Instamatic 126, but the quality of the resulting photos (related to lens clarity, focusing, and exposure control) was substantially superior. The AE-1 was quickly followed by similar products from Japanese camera makers Olympus and Minolta and, in 1984, with an even simpler, less expensive

---

70. For a more skeptical view, see Brock, "Structural Monopoly," who argues that others had pioneered the technological concepts embodied in the 126 and 110 cameras, but were impeded by various Kodak tactics.

71. We ignore here another challenge—a federal court decision that found Kodak in violation of Polaroid's self-developing film patents. As a result, Kodak was required in 1985 to recall all instant cameras it had already sold, exchanging them for disc cameras. Damages of $909 million were assessed in October 1990.

"point and shoot" camera from Canon. These new, higher-quality cameras threatened Kodak in two ways. First, they competed directly with the Kodak Instamatic cameras. In the ten years following Canon's innovation, approximately 13 million standard-format 35-mm cameras, mostly from Japan, were sold in the United States.[72] But second and perhaps more important, those cameras used standard 35-mm film cartridges, readily available from sources other than Kodak, instead of the special-design cartridges required by the Instamatics.

Kodak's direct reaction came in two stages. In 1982, Kodak introduced a completely new pocket-size camera series using film with the individual 8.2 × 10.6 mm exposures arrayed on a disc rather than in the traditional linear format. The product was well received, and by 1986, 23 million Kodak disc cameras had been sold. However, the grainy photos made with disc cameras failed to satisfy the needs of quality-conscious consumers, and Kodak was said to be disappointed as the standard 35-mm format's film market share continued to rise.[73] In 1985, therefore, Kodak reentered the standard-format 35-mm automatic camera market with new K-10 and K-12 models, designed in collaboration with Chinon of Japan and similar to the products of Kodak's Japanese rivals. *Consumer Reports* gave the new K-12 one of its top rankings in 1987, finding inter alia that it had the lowest price of the five top-ranked non-reflex automatic 35s (the other four of which carried Japanese brand names). Further 35-mm product improvements followed. By 1988, Kodak's annual report announced, Kodak had become the U.S. market leader in standard 35-mm camera sales. Almost simultaneously in that same year, emulating the Bic–Gillette history, Kodak and Fuji marketed disposable cameras.[74]

Another challenge from abroad was directed against Kodak's greatest source of profit—film, and especially color film. Fuji of Japan, by far the leading film supplier in its home market, began making inroads into the U.S. market, at first gradually and then aggressively. It started in 1964 with private-label film sales in the United States. In 1972 it began marketing under its own name, first as a give-away with the purchase of Japanese

72. ". . . While Kodak Confronts Doubts about Its 35-mm," *Wall Street Journal,* March 25, 1986, p. 33.

73. See "Kodak Upgrades Disk Film," *Dun's Business Month,* November 1983, p. 25; and "Kodak, Japan's Chinon Plan a 35 mm. Camera," *New York Times,* March 6, 1985, p. D-4.

74. "Playing Leapfrog in Disposable Cameras," *Business Week,* May 1, 1989, p. 34.

cameras, then through sales by specialty camera stores, and in 1978 through broader mass merchandising channels. In 1984, Fuji stunned Kodak by winning sponsorship of the Olympic games in Los Angeles. Fuji mounted a vigorous advertising campaign and announced that its goal was to capture 15 percent of the U.S. film market—an approximate doubling of its 1982 market share.[75] Fuji brought more than aggressive marketing to the fray. In 1977, Fuji introduced a high-speed (ASA 400) color film ahead of Kodak, which followed eight months later. In 1983, both Kodak and Fuji rolled out new faster, sharper-grained color films. Fuji appears to have been first by a week in Japan, but trailed Kodak in the United States.[76] Both firms' products came in ASA 100 and 400 speeds, but Fuji also offered an extremely fast ASA 1600 emulsion that Kodak was unable to match until 1989. From the start of its own-brand marketing in the United States, Fuji's color films yielded richer, brighter colors than their Kodak counterparts. At first Kodak insisted that its colors were what the consumer wanted, but in 1985 Kodak introduced new VR-G 100 and VR-G 400 films which, it claimed, were "the sharpest, most realistic color print films in the world."[77] Fuji retaliated in early 1986 with new Fujicolor Super HR films which, it claimed, provided better color gradation with flash photos. Kodak struck back in 1988 with its new premium-priced Ektar color film series, which embodied important emulsion structure innovations to achieve unusually fine grain without sacrificing speed. In response, Fuji launched its new Reala line in 1989.[78]

Kodak also counterattacked in marketing.[79] It outbid Fuji to obtain sponsorship rights at the 1988 Seoul Olympics, hoping inter alia to blunt Fuji's inroads into the rapidly growing Korean market. To promote its product, Fuji flew its green dirigible over sporting events in Europe and the United States. Kodak retaliated by leasing a blimp in Japan and pho-

---

75. See "Kodak's New Lean and Hungry Look," p. 35; and "Fuji Snaps at Kodak and Hopes It Won't Rear Back," *Marketing and Media Decisions,* July 1984, pp. 32 and 183.

76. "Kodak Unveils 35-mm Films Challenging Japan Firms' Dominance in Their Own Market," *Wall Street Journal,* January 26, 1983, p. 56.

77. Eastman Kodak, *Annual Report: 1985,* p. 6. See also "Kodak Fights Fuji with 'Me-Too' Tactics," *Business Week,* February 23, 1987, p. 138.

78. Compare "Color Is Sharper in New Film Line by Kodak," *New York Times,* October 5, 1988, p. D-1; and "New Kodak and Fuji Films Target Advanced Amateurs," *Wall Street Journal,* March 17, 1989, p. B1.

79. See "The Revenge of Big Yellow," *The Economist,* November 10, 1990, pp. 77–78.

tographing it for its advertisements and New Year's greeting cards with Mount Fuji in the background.[80] In Yokohama, it built an elaborate research facility, and it opened a technical center in Tokyo to provide on-the-spot assistance to laboratories processing Kodak films and papers. And to prevent Fuji from increasing its U.S. sales by acquiring photofinishing laboratories, Kodak brought several such laboratories into its own corporate fold.[81]

Kodak's defense of its traditional market position imposed substantial costs. Its research and development outlays rose from $335 million in 1976, the year the Canon AE-1 appeared, to $1.06 billion in 1986.[82] The average real (inflation-adjusted) rate of increase was 5.6 percent per year. Because of the new price competition it experienced from Fuji, its historically generous film profit margins were squeezed, and overall company profits fell from a peak of $1.2 billion in 1981 to less than $400 million in 1985 and 1986. In the period 1979–1989, its average annual common stock return to investors was slightly below the median for the *Fortune* 500 largest industrial corporations of 1989. In part because of profit shortfalls and in part spurred by observing that Fuji obtained nearly four times more sales per employee, Kodak initiated a cost-cutting campaign in 1983 that, among other things, breached its traditional no-layoff policy. It also brought in outside specialists to train its staff in Japanese-style quality control methods, reducing the incidence of product defects from 32 percent per production lot in 1985 to a planned 10 percent rate in 1987.

Diagnostic Imaging Equipment

Probing the human body non-destructively to detect health anomalies has been revolutionized since the early 1970s by the emergence of several new diagnostic imaging technologies, the most important of which are computed tomography scanners (CT scanners), nuclear magnetic reso-

---

80. "Kodak Fights Fuji with 'Me-Too' Tactics," p. 138.

81. See "Companies That Compete Best," *Fortune*, May 22, 1989, p. 44.

82. R&D effort was also directed toward the more distant threat of electronic imaging as a replacement for traditional film-based photography. See, for example, "Kodak Takes a Risky Leap into Consumer Video," *Business Week*, January 16, 1984, p. 92; and "New Kodak System for Showing Photos on TV," *New York Times*, September 18, 1990, p. C6. But see "Kodak Plans To Cut 3,000 from Payroll," *New York Times*, August 13, 1991, p. D1, in which cutbacks in electronic imaging R&D were announced.

nance imaging (MRI), and ultrasonic scanning. Between 1972 and 1987, U.S. sales of medical imaging equipment (including conventional X-ray devices) grew from $280 million to $2.3 billion.[83]

CT scanning technology originated in the central research and development laboratory of EMI Ltd., a British music recording and electronics firm better known for its enormously successful Beatles records. In 1967, EMI had a policy (motivated in part by the desire to encourage innovations that might yield diversification opportunities) of allowing its laboratories to spend 10 percent of their funds on projects that had not received official budgetary approval. In his work on automatic electronic pattern recognition, the engineer Godfrey Hounsfield observed that conventional film-based methods captured only a small fraction of the information that might potentially be obtained from X-ray beams.[84] Pondering ways of turning his insight to practical use, he conceived an apparatus that scanned its subject with a thin moving X-ray beam, recording the resulting signals through electronic detectors and feeding the information into a computer, which processed the data and produced a three-dimensional representation of the object under investigation. Consultations with British government health authorities led to cooperation between Hounsfield and a neurosurgeon in developing and successfully testing a prototype human head scanner. For his contribution, which was announced publicly in 1972, Hounsfield shared the Nobel Prize in medicine for 1979.

EMI applied for and eventually received patents on the Hounsfield inventions, but it recognized that the basic components of the CT scanner were in the public domain and that other companies were likely to try "inventing around" the patented subject matter. A small EMI subsidiary produced and sold medical electronic devices, but the potential of CT scanning seemed to dwarf EMI's experience in the field. Production of key CT scanner components was therefore subcontracted to outside vendors and two affiliated EMI divisions. EMI asked General Electric in 1971 to distribute its scanners in the U.S. market, which was expected

---

83. See Manuel Trajtenberg, *Economic Analysis of Product Innovation: The Case of CT Scanners* (Cambridge: Harvard University Press, 1990), p. 47.

84. A particularly informative history is Cheryll Barron, "What Scarred EMI's Scanner," *Management Today,* February 1979, pp. 67–76 and 144–150. See also Trajtenberg, *Economic Analysis,* pp. 49–54; and "EMI and the CT Scanner" (A) and (B), Harvard Business School case studies 383-194 and 383-195 (1983).

to account for more than half of the machine's early sales because of America's size and its high propensity to try new medical devices.[85] Evaluating the innovation as technically unpromising and unlikely to lead to sales of more than 30 units, GE declined. EMI thereupon decided to take advantage of its head start and market the scanners on its own in both the United States and continental Europe. The first two U.S. units were installed at the Mayo Clinic and Massachusetts General Hospital in the summer of 1973, leading to the publication in December 1973 of a report on the highly successful results achieved. EMI's order book filled rapidly, stimulated inter alia by EMI's offer of a one-year warranty and its promise to locate a service engineer and spare parts on the customer's premises during the first several months of trial.

At this point, trouble began converging upon EMI from three directions.[86] Internally, EMI had difficulty keeping up with the rapid growth of orders for its original scanner model and developing the new, faster, more versatile models (which, among other things, would scan the whole human torso) that future competition was inevitably going to demand. The transition from prototype to the design of a product that could be manufactured reliably was beset by problems, as were the effort to maintain compatibility between computer software and rapidly changing designs and the coordination of U.K. component production with final assembly at a plant opened by EMI near Chicago. Production backlogs mounted, and in 1977, 90 percent of the scanners already installed in U.S. hospitals were experiencing recurrent breakdowns. Especially for torso scanners, it was clear that scanning times had to be reduced from five minutes, required by EMI's original brain scanner, to a period not longer than patients' ability to hold their breath. A high-speed "fourth generation" design ran into technical difficulties when the required X-ray tubes could not be produced. An alternative design favored by EMI's Chicago engineering team was retarded by technical disagreements with the U.K. central research laboratory staff, and as a result, EMI had lost its technical lead by 1978.

At the same time, as expected, competition began to emerge. Between 1973 and 1978, 14 companies, some traditional X-ray machine makers and some newcomers to the field, attempted to win a foothold as CT

---

85. See Joseph Morone, "GE Medical Systems: A Case Study in the Strategic Use of Technology," manuscript, Rensselaer Polytechnic Institute, November 1990, pp. 5–6.
86. See "EMI and the CT Scanner" (B).

scanner suppliers. One of the first entrants was a small U.S. firm, Technicare (earlier, Ohio Nuclear), whose body scanner was inferior to EMI's counterpart machine. General Electric proved to be a more formidable rival.[87] By the spring of 1974, GE realized that CT scanning was going to be an important market in which it wished to participate strongly. Its first offering was a breast scanner for mammography that emulated EMI's technical concepts and elicited little market interest. In January 1975, it began a "crash program" to develop a rotating fan-beam torso scanner that leapfrogged EMI's pencil-beam scanning technology. The new GE scanner was announced in late 1975. GE had a long-standing reputation among hospitals as a leading supplier of conventional X-ray machines, and with 1,500 field sales and service personnel available, GE was able to provide more intensive technical consultation and repair service than EMI, whose U.S. staff was only one-tenth the size.[88] So strong was customer faith in GE's reputation that 55 of its newly announced fast body scanners were ordered before GE had even begun clinical trials.[89] The trials in 1976 revealed serious image distortion problems, requiring substantial redesign before first deliveries of a still-deficient product commenced in November 1976. Only with the introduction of its greatly improved model 8800 in 1978 did GE achieve technical leadership. Despite some cancellations, the large number of orders booked in advance for GE's early model 7800 and then the successful 8800 significantly eroded EMI sales in the United States. Other early entrants included Siemens of Germany, the world's largest conventional X-ray manufacturer, which distributed Technicare machines until it was able to field devices of its own design in 1977; Picker X-Ray of the United States; and Philips of Holland. Perceiving the difficulties of entering the large Japanese market on its own, EMI in 1974 licensed Toshiba, a Japanese X-ray machine supplier, to assemble and distribute EMI scanners. By 1979, Toshiba had developed its own machines and had begun to market them. Of the 2,058 CT scanners known to have been installed throughout the world between 1973 and 1981, only 28 percent originated from EMI.[90]

The third blow to EMI was a sudden retardation of demand growth in

---

87. For a detailed history, see Morone, "GE Medical Systems."

88. "Can Britain's EMI Stay Ahead in the U.S.?" *Business Week,* April 19, 1976, p. 122.

89. "The Controversy over a Costly Diagnostic Tool," *Business Week,* September 4, 1978, p. 60.

90. Trajtenberg, *Economic Analysis,* p. 52.

EMI's largest single market, the United States. Costing approximately $500,000 excluding installation, CT scanners were among the most expensive equipment items purchased by hospitals. No self-respecting hospital could be without one, but as the diffusion process progressed, a debate broke out over whether smaller, general-care hospitals *ought* to have them. The National Health Planning Act of 1974 required hospitals to file Certificates of Need (CONs) before making major capital investments for which governmental cost reimbursement might be sought. In 1977, regulations were issued that embodied newly elected President Jimmy Carter's campaign pledge to restrict CT scanner placements as one means of curbing escalating medical care costs. New CT scanner installations declined from 208 in 1977 to 122 in 1978 and 141 in 1979 before rebounding in 1981.[91]

With problems in its laboratories and production facilities, from competition, and on the demand side, EMI's Medical Electronics division lurched from a sizable profit in 1977 to equally large losses in 1978. Meanwhile, as the rock and roll boom moved from Liverpool to the United States, profits in EMI's much larger music division also plunged. As a result, EMI was acquired in 1979 by another British electronics company, Thorn Industries, which, to raise cash, sold off the Medical Electronics division. General Electric acquired the division's CT operations outside the United States, thereby becoming the undisputed world leader in CT scanner placements. Antitrust constraints required that the U.S. operations be sold to a smaller firm, Omni Medical, which, faced with increasingly tough competition, failed to sustain EMI's momentum. By 1986, GE's estimated share of CT sales in the United States was 51 percent.[92] Picker (a U.S. company acquired by General Electric PLC of Great Britain in 1981) was second with 19 percent, and Philips of the Netherlands was third with 9 percent. When EMI's U.S. operations were sold to Omni, approximately 25 of its engineers defected to Toshiba, forming a team that developed prototype scanner designs and then transferred them

91. Trajtenberg, *Economic Analysis,* p. 47. In Japan, on the other hand, the Ministry of Health decided in 1978 to make CT scanner costs reimbursible under the national health insurance program. The number of installed scanners thereupon increased rapidly from approximately 200 in 1978 to 6,096 in 1988. Ninety percent of this increased demand was met from Japanese sources. See Fumio Kodama, *Analyzing Japanese High Technologies* (London: Pinter, 1991), pp. 146–149.

92. *Medical and Healthcare Marketplace Guide: 1989* (Philadelphia: MLR Publishing, 1989).

to Japan for production engineering and manufacturing. This helped consolidate Toshiba's position in the second tier of medical imaging device suppliers to the U.S. market.

A technology paralleling CT scanning in many ways is nuclear magnetic resonance imaging (MRI), whose first clinical trials for medical applications began in 1978. MRI, which evolved from instruments used as early as the 1950s for the analysis of chemicals' molecular composition, places the patient in the middle of a powerful magnetic field, causing the nuclei of hydrogen atoms to emit weak signals. The signals are detected and processed by computer to yield images of organs, other tissue, and the like. MRI has two significant advantages over X-ray machines and CT scanners (which also use X-rays). Unlike X-rays, MRI signals are not obstructed by bones, and so MRI can provide excellent images of the spinal cord, brain stem, and other tissues that remain obscured in X-ray analyses. Also, unlike X-rays, MRI's magnetic field is not believed to trigger cancerous cell mutations. MRI's main disadvantages are its inability to work in situations where ferrous metals (such as surgical instruments or clips) are present and its high cost—from one to two million dollars per installation, depending upon the type of magnet structure employed.[93]

It is remarkable, given the problems plaguing EMI's work on CT scanner improvements, that the application of MRI to medical diagnostic imaging was also pioneered at EMI's U.K. laboratories. EMI's research began in 1976 and the first experimental prototype was available in 1978—just when CT scanner sales and profits were evaporating. In the reorganization that followed EMI's acquisition by Thorn, EMI's MRI technology was sold to Picker X-Ray of the United States. Picker's first commercial model was delivered in 1983, propelling the company to a leading position in the new technology. Selling its MRI units through the sales and service organization of its parent, British General Electric, Picker achieved a particularly strong position in Europe.

In the United States, leadership quickly passed to an early CT scanner market entrant, Technicare, which in 1978 was acquired by a giant health-care products manufacturer, Johnson & Johnson. At the time of

---

93. See U.S. Congress, Office of Technology Assessment, Health Technology Case Study 27, *Nuclear Magnetic Resonance Imaging Technology: A Clinical, Industrial, and Policy Analysis* (Washington, D.C., 1984); and "The Hottest Research Is Still Near Absolute Zero," *Business Week,* November 26, 1990, pp. 88–90.

the merger, Johnson & Johnson had negotiated a CT patent license agreement with EMI. Immediately after the merger, Technicare-J&J invested an estimated $5–$6 million in developing its own MRI devices and entered into collaborative research arrangements with 17 universities and medical centers.[94] This collaboration proved to be a mixed blessing, however, because MRI unit placements with collaborating institutions were usually made at unremunerative prices, and nearly half of Technicare's first 44 installations took place under collaborative agreements. By 1986, Technicare had accumulated losses of $260 million under J&J's parentage, and so Technicare's inventories and service commitments on already-installed machines were sold to General Electric (of the United States, not Great Britain).[95] The Technicare manufacturing operations were discontinued.

General Electric was also a late entrant into MRI technology, but once it decided in early 1982 that MRI was technically and commercially attractive, it launched an all-out effort to excel. In 1984, when the first deliveries of its new machines commenced, it employed more than 500 technical staff on its MRI project—more than five times the number so employed at Technicare or at another early MRI entrant, Fonar.[96] Repeating its experience in CT scanning, General Electric drew upon the strength of its reputation and its field organization to book MRI orders a year before its newly developed machines, using powerful superconducting magnets, had undergone clinical testing.[97] In 1987, GE added to its MRI operations by acquiring the medical electronics division of France's Thomson S.A., whose sales were mainly in Europe and South America. By 1986, it made an estimated 37 percent of all MRI equipment sales in the United States, followed by three other U.S.-based companies, Diasonics, Fonar, and Picker, with 22, 12, and 8 percent, respectively, and Siemens of Germany with 7 percent.[98] General Electric claimed in

---

94. "See-Through Investing," *Forbes,* December 21, 1981, p. 32.

95. "Johnson & Johnson to Sell Imaging Line," *Wall Street Journal,* April 10, 1986, p. 7.

96. Office of Technology Assessment, *Nuclear Magnetic Resonance Imaging Technology,* pp. 137–139.

97. In "GE Medical Systems," p. 32, Morone quotes a GE executive's description of his company's come-from-behind MRI strategy: "Blow smoke; develop as fast as you can; publish everything you do; and keep people waiting for whatever it is that GE will do next, so that they don't spend their money on someone else's machine."

98. Adeline B. Hale and Arthur B. Hale, *Medical and Healthcare Marketplace Guide* (Philadelphia: MLR Publishing Co., 1989).

1990 to have more MRI machine placements throughout the world than any rival. But because many companies swarmed into the MRI field once EMI conducted its pioneering demonstrations and because placements were held back by investment restraints on hospitals in the United States and elsewhere,[99] MRI did not yield profits nearly as quickly as CT scanning. GE reported the first profitable year for its MRI operations in 1988.[100]

The second-largest U.S. MRI equipment supplier, Diasonics, was founded in 1977 to manufacture still another type of medical diagnostic equipment, ultrasound systems. Ultrasound operates by sensing the echoes from high-frequency sound waves projected into the object being analyzed—in early applications, aircraft and guided missile weld joints, but increasingly in the 1970s, the human body. Ultrasound is particularly effective in soft tissues and in analyzing circulatory patterns, and it is the only imaging technique able to provide motion pictures rather than "still" snapshots.[101] Ultrasound units are much less expensive than CT scanners and MRI systems, with top-of-the-line models selling for approximately $200,000. Attempting to avoid the difficulty fatal to many other small or inexperienced medical imaging apparatus entrants, Diasonics sought to build up a strong field sales and service force capable of competing on nearly equal terms with General Electric, Picker, Siemens, and other traditional X-ray manufacturers.[102] It moved into MRI in 1981 by purchasing the patent rights from a collaborative research arrangement between the pharmaceutical maker Pfizer and the University of California. By 1983, the same year in which Picker's first EMI-based machine appeared, Diasonics was able to offer a commercial MRI model. However, faulty technology choices in its ultrasound line and in the development of digital X-ray devices led to losses and to the acquisition in 1989 of Diasonics' successful MRI division by Toshiba, which on its own account in 1986 achieved 2 percent of U.S. MRI sales. With that acquisition, four of

---

99. The diffusion of MRI was also impeded in the United States by Food and Drug Administration testing requirements that were not in force when CT scanners were introduced.

100. General Electric, *Annual Report: 1988*.

101. See "Ultrasound Enters New Frontiers," *New York Times*, November 28, 1990, p. D7.

102. As the Diasonics digital radiography president remarked, "Too many start-up companies have the best product, and they can't actually sell it to anybody." See "Diasonics' Winning Ways to Look Inside You," *Fortune*, May 16, 1983, p. 176.

the six leading U.S. MRI suppliers—Diasonics, Picker, Siemens, and Philips—were owned by foreign corporations. The original Diasonics company remained in ultrasonic equipment, sharing U.S. market leadership with Hewlett-Packard; a startup company, Acuson; the Advanced Technology Laboratories Division of Westmark; and Toshiba.

Because many of the organizations selling medical imaging apparatus operate on a multinational basis, some with production in more than one nation and others exporting from a single site, there is a considerable amount of intra-industry trade in the four-digit SIC industry category "X-Ray, Electromedical, and Electrotherapeutic Apparatus" (see Tables 3.2 and 3.3). As the industry was transformed during the 1980s by new technology and mergers and as the dollar rose, at least briefly, in value relative to other currencies, the balance of trade became more even. In 1980, exports were 37 percent of U.S. output, while imports were 17 percent of domestic consumption. By 1986, exports had fallen to 31 percent of output while imports had risen to 30 percent.

Digital Switches

The equipment required to provide modern telephone service is commonly divided into three main classes: switches, which route signals as they arrive at a branching point; transmission gear (for example, cables and microwave radio relay systems), which carry the messages from origin to destination; and terminal equipment, which includes handsets, answering machines, office switchboards, and the like. Central office (CO) switches accounted for roughly half of all telephone operating company equipment outlays during the 1980s.

Central office switches have evolved during the twentieth century from manual coupling of lines across a switchboard by an operator, through automatic connection by moving mechanical wheels or bars, to electronic switches. Electronic switching in turn has two main variants, analog and digital. Analog switching, the earlier of the two to be commercialized, uses electronic devices to allocate a narrow frequency band through which each message is channelled. Digital switching takes messages chopped into digital zero-one "bits" and employs digital computer techniques to assign each bit its correct pathway.

International trade in major telephone company operating equipment has historically been modest, because each national telephone company in the larger industrialized nations tended to favor one or more local suppliers. In the United States, the equipment maker Western Electric was a subsidiary of AT&T, the principal telephone service company until its breakup in January 1984, and AT&T naturally purchased from its own subsidiary when other conditions such as product specifications and cost were roughly equal. In Germany, the telephone operating company is a state enterprise which has tended to purchase from private domestic manufacturers such as Siemens or ITT's German subsidiary, Standard Lorenz. This has been typical of European practice, although a principal goal of the European Community's "1992" campaign has been to break down such national preferences. Until 1985, Nippon Telegraph and Telephone (NTT) was a state enterprise whose policy was to divide its business among several favored domestic manufacturers. Denationalization of NTT did not change matters greatly, despite pressure from non-Japanese nations to have their domestic manufacturers given equal access. But the geographic fragmentation of telecommunications equipment markets did lessen during the 1980s, and that has much to do with patterns of innovation in central office digital switches.

AT&T introduced the world's first large-scale electronic central office switching system, the analog 1-ESS, in 1965. Improved 2-ESS and 3-ESS versions followed, and during the 1970s, AT&T replaced large numbers of old mechanical central office switches with its new electronic analog switches, even though it was recognized that digital switching would eventually come to be the preferred technology. The move by AT&T into digital switches was held back as a result of cost comparisons that failed to project the rapid decline of digital integrated circuit prices.

Research and development on digital CO switches accelerated in various parts of the world during the 1970s. Alcatel of France was the first to introduce an operating system, the E-10A, in 1972. A relatively small producer of mechanical crossbar switching systems, Alcatel joined the electronic era relatively late and decided to bypass analog techniques and jump directly to a digital system. E-10 was adopted by the French national telephone operating company, but export sales were modest, in part because Alcatel recognized that its chances were slim in markets dominated by national champions and partly because the E-10's technical advantages were at best modest relative to existing analog systems. An improved version, the E-10B, introduced in 1979, did not change the situation dramatically. L. M. Ericsson of Sweden was the second to deliver digital CO switches, but it too was limited largely to sales in nations that had no local champion.

In North America, the ice was broken by Northern Telecom (NT), the manufacturing arm of Bell Canada. Until 1956, NT (at the time called Northern Electric) was controlled by Western Electric, AT&T's U.S. manufacturing subsidiary, and served as a producer of equipment designed by Western Electric and its Bell Telephone Laboratories affiliate. After a U.S. antitrust decree required Western Electric to divest itself of equipment sales outside the United States, NT became a subsidiary of Bell Canada (an operating company independent of AT&T). It gradually began to build up its own R&D capability.[103] Between 1967 and 1973, NT produced 1-ESS analog systems under a design license from AT&T, but then NT phased in an analog switch of its own design, the SP-1.

An NT study in 1969 had forecast that sometime between 1974 and 1979, the technical evolution of large-scale integrated circuits would propel digital switching to a position favored over analog. In the early 1970s more detailed design work began. The decision to enter full-scale development, viewed as "the most critical decision NT would make in the 1970s," was taken in 1975.[104] At the time, NT believed that AT&T was planning to produce and install its own digital CO switches in the early to mid-1980s, and other equipment producers were likely to offer digital switches at earlier but unknown dates. The Canadian market was considered too small to justify the large research and development outlays required (estimated in 1975 at $32 million Canadian for a small system and $66 million for a wider range of capacities), and so Northern Telecom planned a vigorous effort to sell its new system outside Canada. (In 1976 it had acquired a plant in Nashville, Tennessee, to manufacture other telephone equipment sold in the United States.) A prototype of its DMS-1 (digital multiplex system) was completed in 1977, and full-scale production commenced in 1979. DMS-1 was designed for central offices with fewer than 12,000 lines, but R&D work continued at a high level on larger-capacity units, the DMS-100 and DMS-200. Both AT&T and the smaller General Telephone (GTE) network lacked comparable products, and so they authorized their operating companies to buy NT digital

---

103. An excellent analytic history is Leonard Waverman, "R&D and Preferred Supplier Relationships: The Growth of Northern Telecom," paper presented at the International Telecommunications Society Conference, Venice, March 1990.

104. See "Northern Telecom" (A), University of Western Ontario case study 9-83-A031 Rev. 10/85 (1982), p. 14. The case study poses the decision-making problem NT faced in 1975.

switches if they desired. DMS sales in Canada and the United States increased rapidly, from $45 million in 1979 to $450 million in 1981.

Meanwhile, Western Electric did not stand still. In 1976 it introduced its first large digital switching system, the 4-ESS, but it was designed to serve as a routing device at major transmission nodes connecting one or more trunklines with others, so it did not fill the central office needs for which the DMS systems were developed. Full-scale development of digital CO switches began in the late 1970s, a prototype was available for test in 1981, and Western Electric began delivering 5-ESS systems to Bell operating companies in 1982. Sales of the 5-ESS grew rapidly, although NT gained at Western Electric's expense from technical difficulties experienced during the first two years of 5-ESS's life. Northern Telecom's fortunes then soared because of a well-timed event anticipated by neither NT nor Western Electric—the divestiture of Bell local operating companies from AT&T under an antitrust case settlement effective January 1, 1984. Under the rules of the divestiture and the new incentive structure, the local operating companies no longer gave preference to Western Electric (renamed AT&T Technologies) when other equipment procurement conditions were roughly equal. Also, large non-communications companies began purchasing digital switches to let their telephone traffic bypass Bell spin-off company central offices altogether. Of a greatly expanded volume of CO switch purchases in the United States during 1984, Northern Telecom won 42 percent, with sales of $700 million to Bell spin-off companies alone. General Telephone's manufacturing subsidiary, which had come up with its own digital switch system in 1982, placed second at 28 percent, while AT&T Technologies ran third at 22 percent.[105] As AT&T Technologies solved its technical problems, NT stumbled in 1985 when two of its DMS installations broke down under call overload conditions. These hurdles too were surmounted, and by 1986, the U.S. central office switch market settled into something approximating a dynamic equilibrium with AT&T Technologies and Northern Telecom each holding shares of roughly 40 percent. Imports did not rise commensurately, however, since NT expanded its manufacturing capacity in the United States so that, by 1988, it had $1.87 billion (U.S.) in plant, equipment, and other identifiable assets in the United States, compared with $1.39

---

105. "Digital Switch Business Sees More Competition," *Wall Street Journal*, January 22, 1986, p. 6.

billion (U.S.) in Canada.[106] Largely on the strength of its successful central office digital switch efforts and similar developments in smaller PBX (private branch exchange) switches, NT's worldwide sales increased from $1.3 billion (U.S.) in 1978 to $4.9 billion (U.S.) in 1987.[107]

The 20 percent of the U.S. CO switch market not captured by NT and AT&T Technologies was the object of a fierce struggle among both domestic and overseas telephone equipment makers, among other things causing prices, according to a trade publication, to come "down, down, down."[108] R&D outlays on NT's DMS-1 program were approximately $76 million (Canadian). Costs escalated into the hundreds of millions for the second generation of higher-capacity switches, and for the third generation, estimates as high as $1 billion have been advanced. Each player in the CO switch development game realized that such large expenditures could scarcely be justified if sales came only in the player's traditional home markets, and until the rewiring of Eastern Europe commenced, most perceived the very large, relatively open U.S. market as the best expansion opportunity.[109]

General Telephone's manufacturing subsidiary, constrained inter alia by the small size of GTE's operating base, failed to sustain strong early sales after the GTE operating companies had largely completed their transition to digital switching. Another U.S. producer, Stromberg-Carlson, had targeted its switches toward small-scale rural applications. It was taken over in 1987 by Plessey of Great Britain, which cancelled Stromberg-Carlson's large-switch development project and focused its efforts, with at least initial success, on selling smaller Stromberg switches to rural Bell operating units. ITT has its corporate seat in the United States, but historically, it sold little telephone equipment in the U.S. market, which was effectively closed through vertical integration. Instead it built strong positions in several European nations and in South America. With the breakup of AT&T, ITT saw an opportunity to achieve what it had long been denied. It refocused the emphasis in the develop-

106. Northern Telecom, *Annual Report: 1987*, p. 31.

107. On NT's aspirations to become the world's leading telecommunications manufacturer, see "Northern Telecom Takes a Big, But Careful, Step," *Business Week*, November 26, 1990, p. 74.

108. "Digital Switching Supplants Other Modes in COs," *Telecommunications*, August 28, 1986, p. 36.

109. See "Siemens, the German Electronics Giant, Belatedly Shoring Up Its U.S. Presence," *Wall Street Journal*, July 19, 1988, p. 24; and "Foreign Firms Struggle to Sell Switches in U.S.," *Wall Street Journal*, October 7, 1988, p. B4.

ment of its System 12 digital line on the special requirements of the United States. With System 12, ITT pursued a strategy somewhat different from that of its international rivals. Extraordinary emphasis was put on developing "distributed processing" software adaptable to a wide range of operating requirements. But ITT experienced major "bugs" in its software programs for U.S. applications, and after expending an estimated $1 billion, it abandoned the entry attempt in 1986, continuing to sell System 12 units in its more traditional markets.[110] In 1987 ITT sold a majority interest in its telecommunications equipment subsidiary to Alcatel's parent, creating a joint venture with Alcatel that held a one-third share of the European market and a strong position in many third world nations.

Other companies merged or entered joint ventures to build strength for the struggle in both the United States and post-1992 Europe. Thus, during the late 1980s, Plessey acquired Stromberg-Carlson; Siemens of Germany acquired General Telephone's European and Taiwan switch-making operations; Ericsson acquired a smaller French equipment producer; Northern Telecom acquired STC of Great Britain; AT&T formed joint ventures with Philips of Holland, which had abandoned an earlier digital switch development effort, and with General Telephone's spun-off U.S. switch development operations; and Nippon Telegraph and Telephone joined Data General Corporation in an effort to develop a high-speed digital communications system for the Japanese and U.S. markets.[111] In the early 1990s, which overseas entity would become the "third supplier" in North America after AT&T Technologies and Northern Telecom remained uncertain.

Fiber Optic Cables

The technological advances in long-distance message transmission have been even more dramatic than those in switching. Originally, one wire

---

110. See "ITT Aims to Sell 25% of U.S. CO Switches," *Telephony,* November 12, 1984, p. 22; "ITT Corporation Steps Up Risky Effort to Sell Digital Switches in the United States," *Wall Street Journal,* May 31, 1985, p. 7; and "For ITT, An Illusory Promise," *New York Times,* June 27, 1986, p. D1.

111. On the delayed development of CO switches in Japan, see Martin Fransman, "Controlled Competition in the Japanese Telecommunications Industry: The Case of Central Office Switches," University of Edinburgh, Institute for Japanese-European Technology Studies, Paper No. 2 (1991).

carried one telephone message. In 1918, the first open wire carrier system was installed by AT&T, permitting a few, and eventually a dozen, messages to be transmitted on a single pair of wires. Coaxial cables carrying 480 telephone signals inside a single thin copper tube buried underground entered Bell System use in 1941. In 1951, coast-to-coast message transmission commenced with a completely different approach, microwave radio relay systems, in which 480 messages were relayed along each of several radio channels from towers spaced roughly 30 miles apart. During the next two decades coaxial cable and microwave relay technology vied for Bell System investment by achieving continuing increases in channel capacity and decreases in cost. As early as 1969, however, in an article describing a new 32,400 circuit coaxial cable, a Bell System magazine predicted that "somewhere in the future, probably before the end of this century, looms still another advance. It is theoretically possible to send messages over a beam of laser light . . . One optical channel . . . would be able to carry the equivalent of more than 10 million voice messages."[112]

Light waves were interesting because they oscillate at much shorter wavelengths, or in other words, at much higher frequencies, than the radio waves used in microwave radio repeaters. And the higher the frequency, information theory had shown, the more bits of information one could superimpose upon the wave form. The future of telecommunications clearly lay in the light spectrum, if the technical problems could be solved.

Basic research during the 1960s revealed that the likely medium for carrying messages astride light waves would be an ultra-pure glass fiber, and the mechanism for generating the carrier light wave would be a laser, whose concept was first articulated in 1958. In the United States, the prime repository of industrial glass technology knowledge was the Corning Glass Works, renamed Corning Inc. in 1989. Bell Telephone Laboratories, jointly owned by AT&T and Western Electric, was a leader in laser technology. During the 1960s and 1970s the two organizations explored and developed the technical possibilities of fiber optic signal transmission, mostly competitively, but with information exchanges and an agreement to cross-license their optical fiber patents. Corning's first demonstration of optical fiber light transmission at attenuation rates (20 decibels per kilometer) attractive for telecommunications took place in 1970.

---

112. George Boehm, "Development Decisions and Expanding Communications," *Bell Telephone Magazine*, March/April 1969, p. 23.

Seeking to free itself from dependence upon AT&T's expertise in cabling and electronics, Corning created a U.S.-based joint venture company, Siecor, with Siemens AG of Germany.[113] Recognizing the difficulty of surmounting national preferences, Corning also sought patents vigorously in other nations and entered patent cross-licensing agreements with several other leading European telecommunications manufacturers, who in exchange agreed not to sell optical fiber or cable made using Corning's technology in the United States.

A key technical milestone was achieved in 1974, when Corning developed a fiber sufficiently pure to transmit signals for 12 miles without amplification—a distance that improved by several times on the amplification spacing requirements of the best available coaxial cables. Dramatic progress continued to be made, but the problems were difficult, and as coaxial cable and microwave technologies continued to advance, the crossover point at which fiber optic cables became more economical than the older methods (and a new competitor, communication satellites) receded in time. As of 1976, Corning's total optical fiber sales following a decade of R&D had amounted to only $1 million. In that year, however, Corning built a pilot plant that could produce 3,000 miles of cable fiber per year, and in 1978, it converted a factory in North Carolina to make optical fibers. Its pilot plant work led inter alia to the development of a process for making fiber cables with only a single continuous strand (that is, "single-mode" cables), providing better long-distance light transmission performance than the multi-strand fibers emphasized previously. Although some Corning executives were skeptical of fiber optics' economic prospects, Corning's television tube business was declining because of Japanese inroads into the U.S. market, replacement sales were desired, and Corning's chief executive officer, Amory Houghton,"felt it was better to be saddled with an idle plant than lose a potential breakthrough after all that investment."[114] The first non-military telecommunications application came when General Telephone installed a seven-mile Siecor network in Long Beach, California, during 1977. AT&T soon responded with a short-haul network in Chicago using its own cable. Five other

---

113. On Corning's joint venture strategy in fiber optics and other fields, see "Corning Incorporated: A Network of Alliances," Harvard Business School case study N9-391-102 (1990).

114. Ira Magaziner and Mark Patinkin, *The Silent War* (New York: Random House, 1989), p. 285. For a more complete history of optical fiber development, see David Chaffee, *The Rewiring of America: The Fiber Optics Revolution* (Boston: Academic Press, 1988).

companies in the United States and numerous others abroad were ac-
tively developing optical fiber cable applications at the time.[115] By 1979,
100 miles of fiber optic cable had been installed for all uses, civilian and
military, in the United States.

Meanwhile, in Japan, rapid progress was being made in this field, as
in so many others. The Japanese government had declared telecommunications to be a priority industry in 1956. State-owned Nippon Telegraph
and Telephone encouraged several of its principal suppliers to plow back
profits on existing contracts into R&D on fiber optics, promising them a
share of the orders it would place when the construction of fiber optics
networks began. A fiber with a transmission loss of 80 decibels per kilometer was demonstrated as early as 1969. Dramatic transmission gains
were achieved during the 1970s by moving to previously unexplored laser
wavelengths. Japanese competition made its first appearance in the
United States in mid-1978, when a nine-kilometer telephone link was
installed at Florida's Disney World, with cabling from Sumitomo Electric
and electronic gear from Nippon Electric.

Japan was the site for the first truly large-scale fiber optics application.
The high population density in the long, narrow island of Honshu provided a particularly propitious environment for cable's high message-
carrying capability. NTT decided jointly with the Japanese government
in the late 1970s to build a new all-fiber "backbone network." Corning
asked to bid on supplying the project's cable needs, but at first was not
given the specifications. Protests through government channels followed,
and when the specifications were turned over, Corning's bid, much lower
in price than those of its Japanese competitors, was rejected because it
used a cable design said to be inconsistent with the specifications.
Whether the rejection was justified is unclear. What is clear is that it
engendered much acrimony.[116]

The early 1980s finally brought the anticipated explosion of optical fiber
cable use in the United States. In 1980, AT&T laid 3,000 miles of cable;
by 1983, its installations had grown to 125,000 miles. Facilitating the

---

115. On the competitive situation in 1978, see "The Fiber-Optics Industry (B): Historical
Development and Competitor Profiles, 1978," Harvard Business School case study 9-379-
139 (1979). A detailed discussion of the technology is found in "The Fiber-Optics Industry
in 1978 (A)," case study 9-379-136 (revised July 1985).

116. See Magaziner and Patinkin, *The Silent War,* p. 289; and Prestowitz, *Trading
Places,* pp. 132–134.

growth of optical fiber cable was the accelerated installation of digital switches, which meshed well with optical fibers' digital transmission mode. An important stimulus was the emergence of long-distance telephone networks, notably, MCI and General Telephone's Sprint, competitive with the AT&T network. As part of their strategy to win customers away from AT&T, the new carriers announced that they would each install transcontinental optical fiber networks. AT&T joined the race, accelerating its own program, and in 1986, it was the first to complete an all-fiber coast-to-coast telephone call. Although AT&T favored its own cable, Corning also benefitted from the competition. AT&T's rivals were anxious not to be dependent upon Western Electric for their network equipment, so in late 1982, MCI ordered 90,000 miles of single-mode Siecor cable (along with electronic components from Fujitsu of Japan). Corning in turn committed $100 million for a new plant capable of producing the cable, which it began delivering in late 1983—the first year its cable venture showed a profit. With the breakup of AT&T in 1984, Siecor was also in a favorable position to compete for the business of the spun-off Bell local operating companies, which were building shorter cable systems within their assigned regions. Smaller amounts of cabling fibers were also sold in the United States by Alcatel of France and several other domestic and foreign companies. During the late 1980s, Siecor supplied from 40 to 50 percent of U.S. demand and AT&T supplied approximately 40 percent.

Japanese companies, led by Sumitomo, entered the 1980s with several times as much optical fiber cable-making capacity as they needed for domestic applications. They began competing for cable contracts in the rapidly growing U.S. market. Despite considerable success in other major industrialized nations, Corning had not been able to obtain patent protection for its cable technology in Japan. When Sumitomo started exporting to the United States at prices considerably lower than those it charged NTT, Corning filed a complaint before the U.S. International Trade Commission charging that Sumitomo's imports infringed Corning's U.S. patents. Sumitomo argued in reply that it had developed its own non-infringing process. The ITC concluded in 1985 that Corning's patents were indeed infringed, but that Sumitomo's 2 percent share of the market was insufficient to find the degree of injury required under the applicable U.S. Trade Act. The battle then moved to the federal courts, with Sumitomo and Corning suing and countersuing over the validity and infringement of Corning's patents.[117] The district court decision in 1987 went to Corning. Sumitomo was required to halt optical fiber production at the plant it had established in North Carolina, and in a negotiated settlement, it paid damages of $25 million to Corning.[118]

Keen international rivalry also materialized over the electronic components required for optical fiber message transmission. In 1982, when AT&T was required to open up procurement for the second stage of its Northeast Corridor cable network to competitive bids, it became known that Japanese engineers were achieving transmission speeds of 400 million bits per second, compared with 45 million bits per second in Western Electric's prototype.[119] A race between technical teams in the two nations ensued, and within a year, Bell Laboratories engineers were able to claim the lead, demonstrating a breadboard model that transmitted 2 billion bits per second over 44 miles without intermediate amplification. Western Electric proved less adept at reducing laboratory designs to reliable production models.[120] One consequence was that AT&T turned abroad for certain critical electronic components, buying from Hitachi many of the lasers for its first transatlantic optical fiber cable, completed in 1988. However, when Fujitsu offered a more advanced electronic design (including Siecor cable) for Bell's Northeast Corridor project in 1982, undercutting Western Electric's $75 million bid by $18 million, AT&T persuaded the Department of Defense to intervene and convince the Federal Communications Commission that national security considerations mandated rejecting the low bid. Western Electric therefore got the job.[121]

During the mid-1980s, the rapid growth of demand for optical fiber cable in the United States ebbed, in part because the first wave of nation-

---

117. For a review of these turbulent events, see U.S. International Trade Commission, *U.S. Global Competitiveness: Optical Fibers, Technology, and Equipment* (ITC Publication 2054, Washington, D.C., 1988).

118. "Corning to Get $25 Million in Suit," *New York Times,* December 6, 1989, p. D4. The basic decision was *Corning Glass Works* v. *Sumitomo Electric Industries, Ltd. et al.,* 671 F. Supp. 1369 (October 1987), amended December 28, 1987, and March 15, 1988. Corning also won patent infringement actions against Sumitomo in Canada, and against ITT (in 1976), a subsidiary of Holland's Philips (in 1981), and two smaller U.S. rivals (in the late 1980s).

119. See Chaffee, *The Rewiring of America,* pp. 36–45, and "Fiber Optics: The Big Move in Communications—and Beyond," *Business Week,* May 21, 1984, pp. 168–182.

120. Chaffee observes, "You can have the best operating transmission system in the world in your own laboratory, but if it's too delicate to operate in the field, what good is it?" *The Rewiring of America,* p. 82.

121. "Japan Runs into America Inc.," *Fortune,* March 22, 1982, pp. 56–61.

wide cable system construction had been completed and in part because improvements in the electronic components permitted greatly increased amounts of information to be transmitted over a given optical fiber channel. The cable side of the business remained dominated by AT&T Technologies and Siecor, while Japanese firms held a substantial position in electronic components. Significant future growth in short-haul applications (including cable television networks) was expected. The Eastern European market, in which communications systems needed to be rebuilt from a small base, appeared particularly promising. Also, important technological advances were expected. For example, switching of telephone, television, and computer signals could in principle be achieved more rapidly with optical switches than with the microelectronic switches that still dominated applications by the early 1990s. With the solution of difficult technical problems in optical switching would come complementary increases in demand for fibers possessing the necessary optical properties. Corning joined forces with IBM and Plessey of the United Kingdom in a cooperative venture, PCO Inc., to pursue opportunities in the new field of "optoelectronics." Following the divestiture of its operating companies, AT&T was forced to cut back its basic research efforts, which had been supported in part by an annual charge on operating subsidiaries.[122] It seemed unlikely, however, that AT&T would curb research in a field as central to its interests as fiber optics.

Facsimile Machines

The idea of transmitting document images (facsimiles) over telecommunications channels dates back to Alexander Baine, a Scottish clockmaker, in 1842.[123] A practical demonstration was achieved at the Chicago World's Fair in 1893,[124] but the first important U.S. application was the Associated Press' Wire-Photo system, inaugurated during the early 1930s. The emergence of heat-sensitive and xerographic copying methods using coated paper during the 1950s freed facsimile communication from

---

122. "Why Own One of the Wonders of the World?" *The Economist*, July 13, 1991, pp. 87–88; and "The Man Who's Running a Nutsier-Boltsier Bell Labs," *Business Week*, August 5, 1991, p. 69.

123. See "Facsimile: Its History, Its Prospects," *Infosystems*, May 1977, pp. 41–42.

124. See "Early Telecommunications Firm Now a Century Old," *The Office*, April 1988, p. 92. Teleautograph Co., successor to the firm making that demonstration, continued to sell facsimile machines, mostly of Japanese design, during the 1980s.

the need for time-consuming photographic development processes at the receiving end and triggered further technological innovations.[125]

Progress in the new medium was also affected by regulatory changes. The most natural pathway for long-distance facsimile communication was over inter-city telephone lines. In the United States, AT&T used tariff restrictions to bar the connection of "foreign devices" into its system, and its Bell Telephone Laboratories and Western Electric subsidiaries were slow to recognize the potential of devices that would compete with their traditional telephone and teletype media. Users of wire-photo and similar equipment were required to transmit their messages at high cost over specially dedicated lines. In the 1960s and early 1970s, a series of legal decisions relaxed and then essentially eliminated AT&T's foreign device connection restrictions. These changes paved the way for non-Bell companies to develop, produce, and sell facsimile devices that could be connected into the Bell System by their users.[126]

As a result of these events, numerous domestic companies, large and small, began producing facsimile devices for the U.S. market between 1964 and 1972.[127] Among the leaders in the early 1970s were Xerox, Magnavox (a radio and television set producer), Minnesota Mining & Manufacturing (3M) (which at first offered Japanese products and then acquired Magnavox's Magnafax subsidiary), Graphic Sciences Inc. (founded in 1967 by two defectors from Xerox and acquired in 1975 by Burroughs), Dacom, and Quip (a new subsidiary launched in a diversification effort by Exxon). By making substantial improvements over the expensive design it inaugurated in 1964, Xerox led the field in 1972, having installed half of the roughly 50,000 facsimile machines used at the time in the United States. Curiously, Xerox's main offerings in that early period used impact printing techniques at the receiving end rather than the coated paper xerographic methods that came into favor later.

---

125. The "electrofax" zinc oxide coated paper process was invented by the predecessor to Xerox Corporation and developed under license by RCA. RCA in turn licensed nearly 100 companies to use its electrofax technology before its principal patent was declared invalid. Thus, unlike the circumstances of plain paper xerography, the technology quickly became widely available. See Erwin Blackstone, "The Copying Machine Industry: A Case Study," Ph.D. dissertation, University of Michigan, 1968, pp. 77–79.

126. A key Federal Communications Commission decision came in the *Carterfone* case, 13 F.C.C. 2nd 420 (1968). See also *Carter* v. *AT&T Co.*, 250 F. Supp. 188 (1966), affirmed in 365 F. 2nd 486 (1966).

127. See "Lots of Talk, Not Enough Fax," *Fortune*, February 1973, pp. 122–134; and "Competition Increases in Facsimile Market," *Industry Week,* September 1, 1975, p. 13.

The diffusion of facsimile technology was held back, however, by a number of obstacles. For one, the early machines were slow and hence entailed high transmission costs. The market-leading Xerox Telecopier 400 took six minutes to transmit a page of information in its high-quality mode. By 1973, however, R&D efforts had identified two ways of reducing transmission times to less than a minute.[128] The machines themselves were also costly; the Telecopier 400 was leased to users at a monthly rental of approximately $600. From these problems and the incompatibility of diverse vendors' machines followed a series of classic chicken-and-egg dilemmas. The demand for machines was limited by the paucity of compatible receiving machines with which would-be senders could communicate and (less important) by the absence of user directories.[129] Most early usage was for internal communication between dispersed units of a given organization, all of which could adopt the same machine. Progress toward standardization was made in 1976, when the Consultative Committee on International Telephony and Telegraphy (CCITT) published standards for so-called three-minute analog machine transmissions. With relatively few users, scale economies and learning-by-doing economies went unexploited and machine costs remained high, as did the costs of special-design integrated circuits, considered the most promising technology for bypassing the portions of a document containing no information and hence reducing transmission costs. And with high costs on the supply side, the broad demand needed to achieve scale economies was not forthcoming.

A breakthrough from this impasse occurred first in Japan, where the demand-side pull was stronger than in the United States. In the United States, teletype machines were a potent competitor to early facsimile systems. But in Japan, with a complex writing system requiring 3,000 different characters, neither teletype nor (at least during the 1970s) computer communications provided an acceptable means of transmitting written messages. Facsimile therefore was accepted more quickly and grew more rapidly. By 1972, after a change in its telecommunications law that allowed facsimile devices to be connected to non-dedicated telephone lines, Japan had approximately 60,000 facsimile machines in operation,

---

128. See Roy Braun, "Outlook on Facsimile Systems: Guarded Optimism Based on Solid Accomplishments," *Infosystems*, July 1973, pp. 40–43.

129. See "Lots of Talk," pp. 126–129, which characterized the situation in 1973 as "building a Tower of Babel."

compared with 50,000 in the United States.[130] The small U.S. firm Dacom led the way to digital signal encoding using integrated circuits, but during the late 1970s Japanese electronics manufacturers pushed the technology aggressively, moving away from the dominant analog (varying tone) method and developing machines capable of transmitting a digitally encoded standard document page in less than a minute.[131]

Even before the new digital technology appeared, facsimile use continued to grow in the United States. By 1977, some 200,000 machines had been installed.[132] Two trends then combined to precipitate explosive growth. Long distance telephone rates fell when MCI and GTE were permitted to establish transmission networks competitive with those of AT&T and when, following AT&T's 1984 breakup under an antitrust decree, the "cross subsidy" traditionally provided to local switching and distribution operations out of long distance charges was sharply reduced. These cost reductions interacted with the drop in document transmission times to a minute (and later to as little as 15 seconds) to make facsimile much more economical, stimulating what proved to be price-elastic demand.

With their new digital machines, the Japanese facsimile manufacturers were well positioned to take advantage of these developments. During the late 1970s they began exporting to the United States and Europe in rapidly increasing quantities attractively priced, standardized high-speed machines. Lacking high-speed machines of comparable quality and cost, some U.S. facsimile pioneers resold machines imported from Japan.[133] Relying entirely upon imports from its Fuji-Xerox affiliate, Xerox retained the largest share of a greatly expanded U.S. market in 1983. The next four positions were held either directly by Japanese suppliers (Matsushita and Ricoh) or by U.S. firms tapping Japanese sources for their high-speed machines (for instance, Oki for 3M and Fujitsu for Burroughs).[134] Sixth-ranked Exxon lost ground in the next two years and eventually exited from facsimile device production. By 1987, the five

---

130. "Lots of Talk," p. 126.

131. The digitally encoded messages could be transmitted over either analog or digital telephone channels. Most of the machines sold through 1988 used analog channels.

132. "Fax Users See Major Changes," *Infosystems*, May 1977, p. 42.

133. "Japan Takes Over in High-Speed Fax," *Business Week*, November 2, 1981, pp. 104–108.

134. *Appliance*, September 1984, p. 67.

front-runners were all Japanese-based companies, with Xerox holding sixth place.[135] Most of the Japanese firms imported their machines to the United States, although Nippon Electric and Murata established U.S. production facilities. By the end of 1988, an estimated two million facsimile machines had been installed in the United States, and 1.7 million more were delivered in the following year.[136] Features such as laser-based plain paper printing appeared in high-end machines, and competition for entry-level machines, largely among Japanese suppliers, drove selling prices to $750.[137] In Europe too, despite French government subsidies to domestic producers,[138] the Japanese achieved a dominant position, supplying more than 90 percent of the facsimile machines sold during 1987.[139]

Why U.S. manufacturers relied upon Japanese sources for high-speed machines in the early 1980s and then failed, with minor exceptions, to develop their own comparable digital models remains somewhat of a mystery. A manager of Ricoh, which later became the leading U.S. facsimile machine supplier, observed in 1981, "Although [the U.S. firms] have the latent ability to develop such machines, they are not doing so, and we don't know why."[140] There were three or perhaps four main reasons. First, with larger and (at least initially) more rapidly growing demand at home, the Japanese producers exploited both static and dynamic scale economies more fully. Their scale-based cost advantage was enhanced by the temporarily high value of the dollar relative to the yen during the early 1980s, when the crucial shift to high-speed machines occurred. Second, it was claimed in 1983 that the Japanese electronics manufacturers combined more fruitfully skills in digital logic, communications technology, and the design of electronic products with costs sufficiently low to tap large price-sensitive markets. One American analyst commented, "If you're a stand-alone facsimile maker like Xerox, I don't see how you

---

135. *Appliance,* September 1987, p. 60; and September 1988, p. 65.

136. "Profiting from Japanese Imports," *The Economist,* April 28, 1990, p. 72; and *Appliance,* September 1990, p. 68. In *Analyzing Japanese High Technologies,* p. 143, Kodama estimates that U.S. facsimile installations totalled 1.1 million in March 1988, compared with 2.2 million in Japan and 600,000 in Europe.

137. With the shift to less expensive console machines and Xerox's loss of market leadership came greatly increased emphasis on machine purchase in place of leasing.

138. "Japan Takes Over in High-Speed Fax," p. 104.

139. "Europe: Switching from the Telex to the Fax," *OEP: Office Equipment and Products,* 16 (December 1987), pp. 36–38.

140. "Japan Takes Over in High-Speed Fax," p. 104.

can survive."[141] This claim is less than fully persuasive, however, given that Xerox had attempted (unsuccessfully) to develop a line of digital computers during the 1970s, Burroughs specialized in digital accounting machines and computers, and AT&T, whose facsimile market share peaked at 4 percent in 1985, combined strong communications and semiconductor skills. A more likely explanation is that the Japanese firms put more emphasis on developing fast, low-price machines to tap what they perceived to be a vast small-business market in Japan. Third, U.S. producers, unlike their Japanese counterparts, underestimated the growth potential of facsimile communication, especially among users of relatively low-priced tabletop machines. This misperception evidently followed in part from the belief that consumers would favor communication between word-processing computer terminals over facsimile communication.[142] Fourth, and related to the third reason, the original U.S. facsimile leaders apparently believed that the best market prospects lay in integrating facsimile machines with entire computerized office automation systems.[143] By focusing too much attention on the whole and not enough on a part for which free-standing demand grew unexpectedly, they conceded to foreign rivals one of the great growth markets of the 1980s.

Earth-Moving Machinery

The earth-moving machinery industry spans a wide array of equipment types, ranging from huge tread-riding crawler bulldozers and mantis-like road graders to small wheeled front-end loaders and backhoes, along with specialized machines such as log skidders, pipe layers, compactors, and off-highway trucks. Some of the American companies that comprise the industry—most notably, the industry leader Caterpillar Inc.—specialize in earth-moving and other construction machinery; others, such as J. I. Case and especially John Deere, also make tractors and related equipment for agricultural applications. Although we shall refer on occasion to the smaller companies, we focus here mainly on Caterpillar, which in

---

141. "U.S. Fax Market Takes Off—with Japanese Engines," *Electronic Business,* November 1983, p. 166.

142. See "Japan Takes Over in High-Speed Fax," p. 108; "Profiting from Japanese Imports," p. 72; and "U.S. Fax Market Takes Off," p. 164. The third (1983) article quotes one "expert" who predicted an imminent decline in facsimile sales, although an Arthur D. Little, Inc., analyst projected continuing 20 percent per year growth.

143. "Japan Takes Over in High-Speed Fax," p. 108.

1981 originated 45 percent of U.S. earth-moving equipment sales and whose story is in many ways the story of the industry.

Into the 1970s, U.S.-based earth-moving machinery companies accounted for approximately 80 percent of free-world production in their various specialty lines.[144] Their early lead was consolidated in the years following World War II, when the manufacturing facilities of West Germany and Japan had been devastated and the U.S. occupation forces left large numbers of "Cat" tractors behind to aid in the reconstruction efforts. Caterpillar built its dominant position on two main principles: sustaining a vigorous program of incremental product improvement to offer machines of the highest quality and reliability, and maintaining an extensive network of largely exclusive parts and service outlets.[145] Product quality and service are important to construction contractors because construction machinery is prone to breaking down under the rigorous conditions to which it is subjected, and machine failure at key stages of a major building project can shut down or slow the entire operation, sapping profits. The more durable and reliable the original equipment design, the less likely such costly breakdowns are. And when breakdowns did occur, Caterpillar, with 106 well-stocked distributorships in the United States and 141 overseas during the late 1970s, assured its customers that it could provide service anywhere in the world within 48 hours 99 percent of the time. Although competitors had gradually eroded its dominant market share, Caterpillar claimed 35 percent of total free-world earth-moving machinery sales in 1981, with more than half occurring outside the United States. Some of Caterpillar's overseas sales were by export from North American plants and some were from its plants in France, the United Kingdom, Brazil, Mexico, and elsewhere. In 1982, 19 percent of its output was produced outside the United States.

The 1970s were boom years for earth-moving machinery makers. A synchronized business cycle peak occurred in most industrialized nations during 1973, and construction activity shared in the prosperity. The dramatic oil price increases effected by OPEC in 1973 and 1974 induced brief recessions in much of the industrialized world, but the slack was more than taken up by massive construction projects initiated by newly oil-rich

---

144. See "A Grim Era for Heavy Equipment," *New York Times,* September 22, 1985, p. 4F.

145. The spirit of Caterpillar was captured humorously in a series of fictional articles featuring Mr. Alexander Botts of "Earthworm Tractor Co." in the *Saturday Evening Post* during the late 1940s.

nations and by the recycling of petrodollars through loans to third world nations, whose infrastructure and industrial investment also boomed. To meet the growing demand, most earth-moving machinery producers greatly expanded their capacity.

The second oil shock of 1979–1981 and ancillary inflation-fighting measures led by the U.S. Federal Reserve in late 1979 caused a dramatic reversal. Europe and, even more, the United States experienced severe recessions, felt with special force in the capital goods and construction industries. At first the OPEC nations gained revenues, but as the demand for their oil declined sharply and then oil prices fell, their construction programs also were cut back. Third world nations were plunged into a debt service crisis, forcing drastic capital program curtailments. In 1985, worldwide construction machinery sales remained depressed at 60 percent of their 1979 levels, and excess capacity abounded.

The U.S. producers were afflicted particularly severely for several additional reasons.[146] For one, high U.S. interest rates during the early 1980s raised the value of the U.S. dollar, putting U.S. companies at a cost disadvantage on their exports and giving foreign equipment makers an advantage in importing to the United States.

In addition, compared with many of the other industries analyzed in this chapter, earth-moving equipment makers' technology is not particularly exotic. What matters most is careful attention to detail and repeated design, field test to failure, and redesign. By the mid-1970s, German, Japanese, Italian, Swedish, and other manufacturers had caught up and were offering machines competitive technically with those of the leading U.S. companies.

Finally, foreign firms began during the boom of the 1970s to invade the U.S. market. In 1976, Fiat of Italy formed a joint venture, Fiat-Allis, with Allis Chalmers of the United States to manufacture an expanded line of crawler and wheel tractors, graders, and the like in Europe, Brazil, and the United States. In 1983, Clark Equipment Company of Michigan took over the bankrupt remnants from the construction machinery operation Daimler-Benz had acquired in 1980 from General Motors. The Clark venture was then merged in a 1985 joint venture with Volvo of Sweden. And in Japan Komatsu, the leading construction equipment maker, was on the move.[147] Through strict cost and quality control efforts, Komatsu had

---

146. See "A Grim Era," p. 4F.
147. See "Komatsu Limited," Harvard Business School case study 9-385-277 (1985).

held a 60 percent share of its home market despite a strong challenge from a 1966 joint venture between Caterpillar and Mitsubishi. During the 1970s it was well positioned to win sales in the booming Middle East and South America, becoming the world's second-largest construction machinery producer. Its initial strategy toward U.S. market penetration was to hold shipping costs down by exporting only relatively compact products such as bulldozers, but by 1981, it had captured a 3 percent share of U.S. construction equipment sales (including 10 percent of bull-dozer sales).[148]

As the dollar rose relative to the yen in the early 1980s, Komatsu became more aggressive. It had adopted as an internal slogan the expression "Maru-C," which means "encircle Caterpillar." Exchange rate swings improved its position in competing with Caterpillar exports to the third world, but more important, Komatsu began vigorously seeking U.S. orders by offering product quality essentially equal to that of Caterpillar at prices 30 to 40 percent below Caterpillar's. By 1985 Komatsu had won 10 percent of U.S. construction equipment orders. At that time, recognizing that a fall in the dollar's value relative to the yen was virtually inevitable and anxious to eliminate its shipping cost disadvantage on bulk-ier machinery, Komatsu began equipping an assembly plant in North Carolina. Komatsu's North American president announced that its near-term U.S. market share goal was 15 percent, with a longer-term goal of 20 to 25 percent.[149]

Caterpillar fought back with an intensity that surprised many who saw it as a giant turned complacently confident of its dominant position.[150]

First, it began discounting its prices—not to parity with Komatsu's, but sufficient to hold its market share, given the price premium its reputation and service advantage would permit. An immediate consequence of this and the general collapse of demand was that it suffered losses in three successive years, 1982 through 1984, cumulating to $953 million.

Second, it commenced a vigorous cost reduction campaign, closing nine plants, cutting capacity by a third, accepting a 204-day strike during 1982 and 1983 in an attempt to resist wage increases, and shedding 40

---

148. See "Komatsu on the Track of the Cat," *Fortune*, April 20, 1981, pp. 164–174.
149. "A Grim Era," p. 4F.
150. For an overview, some details of which differ from the account here, see "Caterpillar-Komatsu in 1986" and "Caterpillar and the Construction Equipment Industry in 1988," Harvard Business School case studies 9-387-095 (1986) and 9-389-097 (1989).

percent of its employees by 1987. Other U.S.-based construction machinery makers acted similarly. Caterpillar moved 4 percent of its production (up from 19 percent in 1982) to lower-cost off-shore factories, increased the automation of domestic plants, and adopted Japanese-like "just in time" production methods to reduce inventory investments and simplify work flows. By the end of 1984, its costs had been reduced by 22 percent relative to 1981 levels.[151] In 1985 it announced the second stage of its cost-cutting program, planning to invest over six years $1.2 billion in a complete reconstruction of its remaining plants, with emphasis on creating facilities that were at the same time highly automated and flexible enough to achieve quick workpiece changes.[152] The ultimate aim was to reduce unit costs by an additional 15 to 25 percent.

Third, Caterpillar revamped its product line strategy. Historically, its strength had been very large, top-of-the-line machines whose comparative advantage was maximized in massive infrastructure and construction projects. New macroeconomic conditions had hit those projects especially hard, and the slump was expected to last indefinitely. Demand shifted to smaller home and office building projects in which the equipment of competitors such as John Deere, J. I. Case, and the like had an advantage. Caterpillar therefore began broadening its line to "the smaller stuff."[153] This it did in part through joint ventures with foreign firms. It obtained from a West German firm three types of track-laying excavators smaller than those it produced internally, as well as four new wheel-type loaders; and it expanded its joint venture with Mitsubishi to include the production of excavators. It also introduced a new "Century" line of small-scale equipment, including a newly developed back-hoe loader that challenged Case's dominant model. Its diversification efforts, through which it planned to achieve half of its sales from small-scale equipment by 1990, were made without increasing R&D expenditures. Indeed, Caterpillar's R&D outlays declined almost continuously from a peak of $376 million in 1982 to $235 million in 1989.

Caterpillar's move toward smaller equipment brought it into market segments in which price competition had typically been stronger than it

151. Caterpillar Tractor Co., *Annual Report: 1984.*

152. See "Cat Is Betting Big on Pint-Size Machines," *Business Week,* November 25, 1985, p. 41; "Caterpillar Rides the Economic Policy Bumps," *Wall Street Journal,* April 5, 1988, p. 37; and "This Cat Is Acting Like a Tiger," *Fortune,* December 19, 1988, p. 69.

153. "This Cat Is Acting Like a Tiger," p. 69.

was for top-of-the-line machines. The dollar's fall during the second half of the 1980s and the partial revival of demand helped Caterpillar achieve record profits of $616 million in 1988, declining to $497 million in 1989. The stock market reacted with restrained enthusiasm. Caterpillar's common stock traded in the range of $53–$69 per share during the boom of 1989, compared with $43–$62 a decade earlier, when the *New York Times* stock price index stood at one-third its 1989 average. Caterpillar's chief executive officer dismissed Wall Street analysts' concerns about the cost of the company's modernization program, arguing that "we are absolutely convinced of the payoff in the long term . . . We have to play the long-term game because the competition is playing the long-term game."[154]

While benefitting Caterpillar and other American producers, the increase in the yen's value squeezed Komatsu. Between 1985 and 1988, Komatsu was forced to raise its prices seven times by a total of 40 percent. Komatsu's U.S. president observed publicly in 1986, "If they [Caterpillar] maintain their current prices, I think we will lose market share . . . I hope that they raise prices."[155] Caterpillar in fact raised its prices at much lower rates than did Komatsu, recapturing one-fourth of Komatsu's U.S. market share by the end of 1988. Komatsu responded by reorganizing its world production network, moving more of its production away from Japan, increasing its user service efforts, developing new products to tap specialized construction needs, diversifying into plastics, and putting more emphasis on its metal-forming machinery and industrial equipment operations. And perhaps most important, it surrendered a part of its autonomy. In 1988 it combined its own Western Hemisphere operations in a 50-50 joint venture with those of Dresser Industries, which in 1982 had purchased the construction machinery operations of the bankrupt International Harvester. Dresser added to them in 1984 the Westinghouse Air Brake Company off-highway truck line. The combination, which became the second-largest construction machinery producer in

---

154. "Thinking Long Term Is Costly to Caterpillar," *New York Times,* November 24, 1989, p. D1. See also "Cat vs. Labor: Hardhats, Anyone?" *Business Week,* August 26, 1991, p. 48.

155. "Caterpillar: A Test of U.S. Trade Policy," *Los Angeles Times,* June 8, 1986, p. IV-6. See also "A Weakened Komatsu Tries to Come Back Swinging," *Business Week,* February 28, 1988, p. 48; and "This Cat Is Acting Like a Tiger," p. 72.

North America, experienced severe cultural mismatch and coordination problems, and in 1991 its prospects remained uncertain.[156]

Conclusion

U.S. manufacturers have reacted in widely differing ways to new technological challenges from abroad. Some, such as Gillette, Eastman Kodak, General Electric (in diagnostic imaging), and Corning-AT&T (in fiber optics), redoubled their own R&D efforts in response to growing high-technology competition. Others, notably the makers of such consumer electronic apparatus as television sets and video recorders and the early market leaders in facsimile machines, in effect surrendered the market to overseas rivals. Still others, such as the calculator and tire manufacturers, AT&T Technologies (in digital switches), and Boeing, continued vigorous R&D efforts but ceded sizable market niches.

There are equally diverse reasons why domestic market leaders found themselves behind technologically. In an important change from the factual circumstances that originally inspired the product cycle theory,[157] the U.S. market no longer has clear pride of place as the most fertile ground for pioneering new product developments owing to its superior size and prosperity. Japanese firms developed facsimile machines aggressively in part because facsimile was a preferred means of transmitting documents written in complex Kanji characters. Honshu's geographic configuration and high population density made Japan a logical first locus for a fiber optic cable network. Unique driving conditions nominated Europe as the natural origin of steel-belted radial tires. Nor are the first movers in a new technology necessarily constrained by the size of their home market. In many industries, a world view has taken hold, and sales abroad follow close on the heels of sales at home. A transnational perspective clearly contributed to Northern Telecom's pioneering in digital switches, EMI's development of CT and MRI scanners, the introduction of stainless steel razor blades and disposable razors, the Airbus venture, and the development of video cassette recorders and compact disc audio players.

The locus of technological innovation is also affected by the presence

---

156. "A Dream Marriage Turns Nightmarish," *Business Week*, April 29, 1991, pp. 94–95.

157. See Chapter 2.

of special technological skills, for instance, when production of the first electronic calculators migrated to Japan because well-honed electronic systems assembly skills and low labor costs were found there. Further honing of those skills facilitated the development of VCRs and high-speed facsimile machines in Japan. Cost advantages can also emerge from the vagaries of exchange rate movements. The most dramatic Japanese inroads into the U.S. market for high-speed facsimile machines, hand-held calculators, and earth-moving machinery coincided with periods when the Japanese yen exchanged at relatively low rates against the dollar. Government subsidies were instrumental in Airbus Industries' development programs and contributed to the Japanese firms' strong performance in optical fiber cabling and electronics. Finally, serendipity affects the pattern of events. The knowledge needed to pioneer CT scanning existed in several nations, but EMI's Hounsfield appeared with the right idea at the right time. First technological moves rooted in ground fertile for any of these reasons are often reinforced by early cost reductions achieved through learning by doing, complemented by aggressive pricing to exploit and extend one's cost advantage. Japan's successes in VCRs and facsimile machines are particularly clear examples.

Companies well established in their American home markets have left themselves exposed to technological challenges from abroad in numerous ways. Complacency has been a significant source of vulnerability. Eastman Kodak needed a shock from Fuji to realize that its powerful reputation and entrenched domestic position did not compensate for inferior camera quality and color films that failed to satisfy consumer preferences. The tardiness of U.S. television manufacturers in adopting solid-state color technology stemmed in part from complacency, aggravated by excessive reliance upon RCA as a source of innovations.[158] Despite high expenditures on research and development, RCA stumbled badly in television sets and video recorders because its management system became more adept at projecting returns on investment than shepherding innovations from the laboratory to the factory.[159] Superior patience in attending to technical details and transforming designs into low-cost but reliably

---

158. Our finding on U.S. photographic and consumer electronics firms' vulnerability parallels Michael E. Porter's conclusion, from studies of other industries, that strength in international competition follows from vigorous domestic rivalry. *The Competitive Advantage of Nations* (New York: Free Press, 1990), especially pp. 117–122.

159. See, for example, Sobel, *RCA*, especially pp. 157–161 and 164–167.

producible hardware also contributed to Japan's success in VCRs, facsimile machines, and fiber optics electronics.[160] The introduction of stainless steel razor blades and, less clearly, radial-belted tires and digital switches had to await foreign firms' moves as a consequence of U.S. market leaders' reluctance to cannibalize profits derived from sales (or, for digital switches, from use) of products embodying earlier technology vintages. Lags in recognizing important technological opportunities also played a role, especially in facsimile machines, central office switches, and diagnostic imaging systems. Overseas-based innovators' entry was invited in some cases by the failure of domestic firms to cover all significant product market niches—for example, in color film, miniature television sets, and wide-body shorter-range jet airliners. Patent protection appears, at least for our sample of industries, to have provided only weak barriers to technological inroads by foreign rivals, and in television, RCA's patent licensing policy helped Japanese companies secure a foothold upon which further gains were built. The principal exception was Corning's use of its patent portfolio to drive Sumitomo from the U.S. market.

Although U.S. firms have sometimes been caught napping by transnational technological competition, in more cases than not, at least within our sample, the leading domestic incumbents recognized that they had much to lose by failing to respond aggressively. The most common response was an intensification of product and process development efforts. Having widely dispersed multinational operations helped firms avoid being surprised, as foreign branches in effect served as a "distant early warning line." Gillette's early recognition of the disposable razor threat, the experience Goodyear and Uniroyal accumulated producing radial tires in Europe, Kodak's contacts with Fuji in Japan, and Caterpillar's competition with Komatsu in the third world provide examples. In contrast, however, Xerox's Fuji-Xerox affiliation did not prevent a serious loss of U.S. facsimile market share when other Japanese firms intensified their assault during the early 1980s.

Aggressive reactions on the pricing dimension were less common. The principal exceptions were Caterpillar and Eastman Kodak, both of whom cut their own prices to meet price competition from high-technology imports and simultaneously undertook strenuous cost-cutting campaigns. It

---

160. On the importance of detailed problem-solving in design and early production, see Dertouzos et al., *Made in America,* especially chap. 5.

is noteworthy that the stock market reacted negatively to their sacrifice of short-run profits to maintain long-run market position. Corning too appears to have been penalized by the market for its far-sighted investments in the potential of fiber optics. It is at least arguable that Texas Instruments chose a preferred course in abandoning the price-competitive market for simple calculators and concentrating on niche demands for more complex programmable devices. When such a niche strategy is pursued, or when domestic producers react to tough import competition by giving up altogether, as in video recorders and facsimile machines, the United States cedes a part of the technological frontier to other nations, and international trade in high-technology products becomes more evenly balanced than it was during the 1950s and 1960s.

Marketing channel control figured prominently in some firms' defense against high-technology import competition. General Electric's strong comeback after EMI took the lead in medical imaging devices was attributable in part to GE's superior hospital market sales and service network, permitting it to win and hold customer orders while its late but "leap-frogging" products were still under development. Caterpillar's defense against Komatsu was also bolstered by a strong field service organization and a reputation for quick spare parts supply. But well-established distribution channels were insufficient to protect domestic mechanical calculator and low-speed facsimile machine makers from more innovative Japanese offerings, because the newer substitutes were sufficiently reliable to reach consumers through low-service outlets.

When other defenses fail, domestic producers frequently turn to the government for protection. Television set manufacturing is the most important example on the United States market side of our case study sample.[161] The orderly marketing agreements initiated in 1977 failed to induce behavioral changes sufficient to reverse the tide, and in the end, most of the U.S. television set producers became subsidiaries of companies with a home base abroad. Corning's plea for import restraints under the U.S. Trade Act was unsuccessful, although conventional patent litigation ultimately had a more dramatic impact in halting Sumitomo's inroads. AT&T pleaded jeopardy to national security to defeat Fujitsu's low bid as a

---

161. There were also important barriers to U.S. firms' sales in overseas markets—for instance, in the early closure of the Japanese television set market, in the rejection of Corning's bid to supply optical fiber cable for the Honshu backbone project, and in the preferences overseas telephone companies have exhibited for domestically produced switches.

would-be supplier of optical fiber cable systems for New England. U.S. government restraints on hospitals' CT scanner purchases in 1977 and 1978 gave General Electric valuable time to overtake EMI technologically. But government protection is not only given; it can be taken away. The divestiture of AT&T operating companies and the deregulation of long-distance telecommunications entry opened important opportunities to domestic and foreign competitors—for Corning in optical fiber cables, and for Northern Telecom and others in central office digital switches. Invigorated technological competition followed.

4   Imports, Exports, and
Intra-Industry Trade

Case studies provide important qualitative insights, but they are difficult to generalize. To obtain a more comprehensive perspective on the trade patterns and competitive interactions that emerged in U.S. manufacturing industries during recent decades, we advance to the quantitative phase of our analysis. In this chapter we provide more detailed statistics on import and export trends and investigate the magnitude and determinants of intra-industry trade. Chapter 5 then examines how the research and development spending of U.S. companies responded to generally increasing high-technology import competition.

Industry Import and Export Data

Essential to our analysis are statistics on imports and exports, comparable over time and finely disaggregated to the level of individual manufacturing industries. In this chapter and the next we use data on 449 four-digit SIC (Standard Industrial Classification) manufacturing industries for the years 1963 through 1987. The data were derived after considerable "cleansing" and extensions from a series compiled at the National Bureau of Economic Research.[1] Details of the data compilation process are discussed in the appendix to this chapter.

Figure 4.1 plots trends in imports as a percentage of domestic manufactured goods output value for three diversely compiled time series: the simple average of the import/output percentages across 449 manufacturing industries (dot-dash line), the same ratios derived from trade statistics

---

1. We are grateful to John Abowd, who prepared the data, and Larry Katz, who provided the data in machine-readable form.

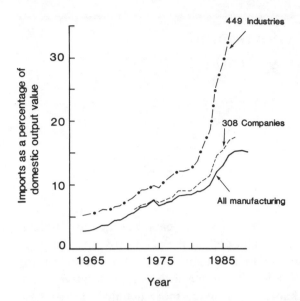

Figure 4.1   Trends in U.S. manufactured goods imports

aggregated to the all-manufacturing level (solid line), and the simple average of sales-weighted average industry import/output percentages (broken line) for 308 relatively large and technology-intensive companies that will be the focus of analysis in Chapter 5. Figure 4.2 presents export/output value percentages from the same three sources.

Several features are noteworthy. First, the general trend was toward increasing U.S. foreign trade. Second, tight money and the relatively high value of the U.S. dollar during the early 1980s stimulated a surge of imports, while exports departed from their previous trend and declined. Third, import/output ratios averaged over 449 individual manufacturing industries are considerably higher than the aggregate all-manufacturing values and the averages for 308 companies. There are two reasons for this: very high ratios of imports to domestic output (well over 100 percent) were recorded in a few industries,[2] and the figures for those indus-

---

2. Thus, if two-thirds of domestic demand is satisfied through imports and one-third through domestic production (and assuming no exports), imports are 100 (66.67/33.33) = 200 percent of domestic production. Import percentages exceeding 100 are commonly avoided by measuring imports as a percent of domestic consumption (that is, imports/

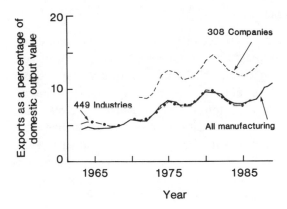

Figure 4.2   Trends in U.S. manufactured goods exports

tries, and more generally for small industries with relatively high import ratios, receive more weight in simple than in value-weighted averages. Differences of this character are much less evident for the export figures. Fourth, the members of our 308 company sample had considerably higher weighted average export ratios than the all-manufacturing universe. This is so because the companies had home bases in relatively high-technology industries, which, we saw in Chapter 2, have enjoyed comparative advantage in exporting. Finally, weighted average import ratios across industries served by the 308 companies rose somewhat more rapidly than the aggregate import values for all manufacturing enterprises. This reflects the increasing prominence of high-technology imports in U.S. markets, especially during the 1980s, when the correlation between net exports (exports minus imports) as a percentage of industry output value and industry research and development intensity deteriorated.[3]

(output plus imports minus exports). But computation of net export ratios on a gross national product accounting basis, as we shall do in the next chapter, mandated the use of domestic output value as a common denominator.

Import/output percentages exceeded 100 in four of the 449 industries for 1975, in 11 for 1980, and in 23 for 1985. In 1985 (but not in other quinquennial years) the percentages exceeded 1000 in three industries—miscellaneous non-rubber shoes, fine earthenware food utensils, and jewelers' materials (SICs 3149, 3263, and 3915). None of the 308 companies in our Chapter 5 sample had an appreciable position in the three most import-impacted industries.

3. See Table 2.1.

Measuring Intra-Industry Trade

We observed in Chapter 2 that differences in home market product characteristic preferences and the quest for economies of scale may give rise to intra-industry trade. For instance, U.S. manufacturers may export microprocessor semiconductors and simultaneously import dynamic random-access memory chips.[4] We attempt now to measure the phenomenon and to identify statistically the conditions under which it thrives.

Some economists have argued that the concept of intra-industry trade is merely definitional because, if "industries" are defined narrowly enough, the phenomenon disappears. Thus, one would not expect to see many newly manufactured Leica cameras exported from the United States to Germany. And new Honda Accord sedans moved only from Marysville, Ohio (where their production was solely located in 1990), to Japan and other points, not in the reverse direction. But this criticism goes too far. Although it may be difficult to define with precision what an "industry" is, any reasonable construction of the concept will group together in one category firms using more or less similar technologies to produce goods with broadly similar characteristics, even when the products are not perfect substitutes. There is no "Honda compact car" industry or "Leica" industry. Because the 449-fold four-digit Standard Industrial Classification breakdown used here comes tolerably close to defining "industries" meaningfully, we linger no more on semantic quibbles.

Given appropriate industry definitions and trade data commensurately disaggregated to the industry level, there remains the problem of finding a measure that captures the essence of intra-industry trade. The measure used most commonly by economists is the Grubel-Lloyd index, defined as:[5]

$$[4.1] \quad IIT_{GL} = \frac{(\text{Exports} + \text{Imports}) - |\text{Exports} - \text{Imports}|}{\text{Exports} + \text{Imports}},$$

---

4. The phenomenon was first given serious attention when Bela Balassa observed that, following the reduction of tariffs by European Economic Community member nations, the emerging trade patterns were characterized much more by cross-shipment of goods in the same industry categories than concentration of any given industry's export activity in the manufacturers of a single member nation. See Bela Balassa, "Tariff Reductions and Trade in Manufactures among the Industrial Countries," *American Economic Review,* 56 (June 1966), pp. 466–473.

5. Herbert G. Grubel and P. J. Lloyd, "The Empirical Measurement of Intra-Industry Trade," *Economic Record,* 47 (December 1971), pp. 494–517. For a discussion of other, less widely used, indices, see David Greenaway and Chris Milner, *The Economics of Intra-Industry Trade* (Oxford: Basil Blackwell, 1986), pp. 60–71.

where the vertical bars enclosing the (Exports − Imports) term denote absolute values, with algebraic signs ignored. The Grubel-Lloyd index ranges in value from 0, implying no intra-industry trade, to a maximum of 1.0. A significant disadvantage is that the index characterizes the *balance* between exports and imports more effectively than it measures the *levels* attained by both. Thus, if imports are 1.5 percent of output and exports are also 1.5 percent, the Grubel-Lloyd index attains its maximum value of 1.0, even though one can hardly say that trade is very active in either direction.

The problem here is that the notion of intra-industry trade involves two potentially conflicting phenomena—substantial amounts of trade and a relatively equal balance between industry exports and imports. To capture these two properties simultaneously, we have devised a new index:[6]

$$[4.2] \qquad IIT_{SH} = \frac{(\text{Exports} + \text{Imports})/2}{\text{Standard Deviation of Exports and Imports}},$$

whose numerator is simply the average *level* of imports and exports as a percentage of output value in a given year and whose denominator measures the extent to which the two magnitudes *differed*.[7] Thus, IIT_{SH} rises with greater trade and also with less disparity between imports and exports.[8] One difficulty is that as the import and export ratios approach equality, the denominator approaches zero, causing the index to explode. To avoid the problem, the standard deviation was replaced by a fixed cutoff value (set at 3.5 in most analyses that follow) whenever it fell

---

6. The "*H*" subscript is for Keun Huh, the author's research associate at the U.S. Census Bureau in 1990, who participated in the index's genesis.

7. If X is exports and I imports, the standard deviation is calculated as

$$\sqrt{[X - (X + I)/2]^2 + [I - (X + I)/2]^2},$$

assuming only one residual degree of freedom. In this special case, the calculation reduces simply to $0.7071 \, |X - I|$.

8. When, for 449 industries, IIT_{SH} indices for 1975 were regressed on their two component parts *NUM* (for numerator) and *DEN* (for denominator), the resulting fitted equation was

$$IIT_{75} = .97 + .172\,NUM - .158\,DEN, \qquad R^2 = 0.682,$$
$$\phantom{IIT_{75} = .97 + .1}(30.87) \qquad (29.18)$$

with *t*-ratios given in subscripted parentheses. Although the zero-order correlations between the index and its components are small (0.275 for *NUM* and −0.056 for *DEN*), each component makes a significant and relatively equal contribution. The regression would be singular if estimated in the logarithms.

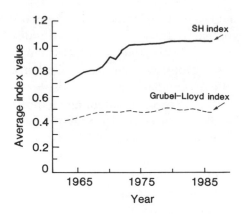

Figure 4.3   Trends in U.S. intra-industry trade, averaged over 449
             manufacturing industries

below that value. The choice of a cutoff value is essentially a choice concerning how small a difference in the export and import percentages can be before it is no longer meaningful. The chosen value of 3.5 implies that differences smaller than about five percentage points do not reflect meaningful import and export flow disparities for purposes of judging whether trade is well balanced. Sensitivity analyses of alternative cutoff choices will be reported.

Figure 4.3 plots the simple average values of IIT_{SH} across 449 four-digit manufacturing industries for the years 1963–1986 (solid line). Also shown are annual average values of the Grubel-Lloyd index (broken line). Both indices exhibit an upward trend, but the trend is stronger, especially before 1973, for IIT_{SH}, whose average value doubled from 0.704 in 1963 to 1.411 in 1986. In addition to having somewhat different average trends, the two indices are not highly correlated cross-sectionally with one another in individual years. Thus, the simple correlation between them was 0.349 for 449 industries in 1965, 0.496 for 1975, and 0.536 for 1985.

That the IIT_{SH} index captures more effectively what is meant by intra-industry trade is suggested by Table 4.1. It tallies 1975 exports and imports as a percentage of domestic output value, the new index, and the Grubel-Lloyd index for three groups of industries—those with the seven highest IIT_{SH} values within a 272 industry subsample that will be emphasized in subsequent analyses; seven industries clustered about the median value of the new index; and the seven industries with the lowest non-zero IIT_{SH} index values. For the top-ranked industries, both the Grubel-Lloyd index and our new index have high values. From the import and export ratios, it is clear that industries with both a considerable volume of trade and relatively evenly balanced trade have been selected. For the lowest-ranked industries, the story is quite different. Four of the seven Grubel-Lloyd indices have above-average values due to the close similarity of export and import ratios, despite the absolutely small trade volumes. The median-ranked indices received middling IIT_{SH} index values because they

Table 4.1 Intra-industry trade statistics for top-ranked, median-ranked, and low-ranked industries, 1975

SIC code (1982 basis)	1975 exports as % of output	1975 imports as % of output	IIT_{SH} index[a]	Grubel-Lloyd index[b]
Top-ranked industries				
2611	39.7	43.6	11.90	0.95
3552	35.3	35.3	10.08	1.00
3693	26.4	23.4	7.11	0.94
3699	26.4	21.6	6.86	0.90
3629	19.8	21.6	5.91	0.96
3832	14.4	16.6	4.44	0.93
3541	17.4	12.7	4.29	0.84
Median industries				
3497	2.4	5.8	1.17	0.59
3567	28.9	7.0	1.16	0.39
3651	11.0	50.0	1.11	0.36
3952	4.6	3.2	1.10	0.82
3873	8.2	38.6	1.09	0.35
3951	18.7	3.9	1.08	0.35
2899	22.9	4.8	1.08	0.35
Low-ranked industries				
2647	0.7	0.6	0.19	0.94
2648	0.6	0.6	0.17	0.99
2655	0.5	0.6	0.16	0.94
3761	1.0	0.0	0.15	0.00
2652	0.9	0.1	0.14	0.16
3411	0.4	0.3	0.11	0.84
2651	0.6	0.1	0.09	0.25

a. Equation [4.2], with a minimum standard deviation of 3.5.
b. Equation [4.1].

had either large volumes of trade with considerable inequality or intermediate but well-balanced trade volumes. The Grubel-Lloyd index assigns relatively high index values to the latter but not the former.

Hypotheses and Explanatory Variables

We advance now to the question, why do observed levels of intraindustry trade vary so much from industry to industry? Among the causal hypotheses enumerated in an admirable survey by David Greenaway and Chris Milner, six are particularly relevant to an analysis of industry cross-section data for a relatively highly developed nation like the United States.[9] Specifically, intra-industry trade is expected to be greater:

1. the greater the potential for physical differentiation of industry products (for example, through new product development);
2. the greater the scope for economies of scale along any or all of three dimensions: static; dynamic, through learning by doing; and through the necessity of amortizing large front-end R&D costs;
3. the more industry structures approximate conditions of monopolistic competition (that is, with differentiated products and open entry);
4. in internationally oligopolistic industry structures;
5. when tariff barriers are low; and
6. when there are relatively high levels of foreign direct investment.

Despite a considerable amount of prior empirical work,[10] many of these hypotheses have not been subjected to clear-cut statistical tests because relevant, well-measured explanatory variables have been unavailable. Here we are able to proceed farther than earlier studies thanks to a remarkable survey carried out at Yale University during the early 1980s.[11] Detailed questionnaires were completed by 650 U.S. industrial research and development managers, providing evaluations over an array of attri-

---

9. *The Economics of Intra-Industry Trade*, pp. 110–123. See also Bela Balassa and Luc Bauwens, *Changing Trade Patterns in Manufactured Goods: An Econometric Investigation* (Amsterdam: North-Holland, 1988), chaps. 4 and 5.

10. A comprehensive survey is found in Greenaway and Milner, *The Economics of Intra-Industry Trade*, chap. 9.

11. See Richard C. Levin, Alvin K. Klevorick, Richard R. Nelson, and Sidney G. Winter, "Appropriating the Returns from Industrial Research and Development," *Brookings Papers on Economic Activity*, 1987, no. 3, pp. 783–820.

butes such as the relative importance of various means by which their organizations captured the benefits from product and process innovations; the relevance of company links to basic science and academic engineering research; the rate at which technological change had been occurring, and was expected to occur, in their industries; the objectives of company R&D programs; and much else.

The R&D executives' responses pertained to 130 industries, mostly defined at the four-digit SIC or broader three-digit level, and mostly with a relatively strong emphasis on research, development, and technological innovation. In order to eliminate weak coverage and mismatch problems and also for reasons that will be elaborated in the next chapter, we reduced the universe of 449 four-digit manufacturing industries to a subsample of 272 four-digit industries meshing more closely with the Yale survey responses. Excluded were all industries in the characteristically low-tech two-digit SIC food, tobacco, textile, clothing, wood products, printing and publishing, and leather goods groups. Also excluded were the petroleum refining industries, whose import patterns were distorted by oil shocks and government controls between 1973 and 1981. Of the 130 industry categories covered by the Yale survey, 105 pertain to the 272 industries retained in our subsample. When the Yale industry categories were broader than four-digit SIC industry definitions, the responses for the broader categories were repeated for each constituent four-digit industry. This introduces some measurement error. The average number of R&D executive responses elicited per industry category was five, but in 30 cases, there was only one response per category. The sparsity of responses in some cases and the unavoidable problem of perceptual error in complex survey responses again imply significant errors of measurement in our explanatory variables, imparting a bias toward zero in estimated regression coefficients.

The Yale survey permits extraordinarily rich tests of the role product differentiation plays in intra-industry trade. From it, we identify four particularly relevant variables: *PROGRESS*, measuring the rate at which new and improved products were introduced in the respondents' industries during the 1970s; *SCIENCE*, assessing the relevance of the basic sciences of physics, biology, and chemistry (average of three scores) to industry technological progress; *ENGINEERING*, assessing the relevance of university-based research in chemical, electrical, and mechanical engineering (average of three scores); and *STANDARD*, asking the extent to which an industry's technical activities were directed toward achieving standardized or dominant product designs. The first three are expected to be positively associated with high levels of intra-industry trade. Emphasis on standardization, in contrast, tends to coincide with late product life cycle stages, and might therefore be expected to have a negative impact on intra-industry trade. Two further product differentiation variables are drawn from Federal Trade Commission Line of Business surveys.[12] *RD/S* measures company-financed research and development outlays as a percentage of sales, averaged over the years 1976 and 1977. *ADV/S* measures media advertising outlays as a percentage of sales in 1977. The product innovation activity that gives rise to intra-industry trade is expected to increase with *RD/S*. *ADV/S* has been used in prior studies as an index of trade-stimulating product differentiation, although it is questionable a priori whether the "image" differentiation emphasized through heavy media advertising is in fact a solid basis for trade.

Three variables take into account various aspects of trade-enhancing scale economies. Front-end fixed costs may be larger in industries pursuing research and development intensively, so higher *RD/S* values again imply greater intra-industry trade. The Yale survey asked respondents how important first-mover advantages embodied in learning by doing were in appropriating the benefits from new products. We call the resulting variable *LEARNING*, expecting a positive association with intra-industry trade. The concentration of industry sales in the hands of the largest sellers tends to be greater in industries requiring sizable plants to achieve all static economies of scale.[13] We therefore include the variable *CR4*, measuring for each four-digit industry the combined U.S. output value share of the four leading domestic manufacturing producers during 1977.

The concentration ratio *CR4* also characterizes domestic market structures, indicating whether they are atomistic, oligopolistic, or monopolized. Because intra-industry trade is believed to thrive in oligopolistic

---

12. See Federal Trade Commission, Bureau of Economics, *Statistical Report: Annual Line of Business Report* (Washington, D.C., May 1982 [for the 1976 report] and April 1985 [for the 1977 report]). The R&D variables are averages of 1976 and 1977 values when both were available, or from whichever year was available for industries on which data for only one year were published.

13. See F. M. Scherer, Alan Beckenstein, Erich Kaufer, and R. D. Murphy, *The Economics of Multi-Plant Operation: An International Comparisons Analysis* (Cambridge: Harvard University Press, 1975), p. 194, where the correlation between an index of relative plant size and the four-firm concentration ratio is found to be +0.76.

structures, we include a quadratic term *CR4SQ* to capture possible non-linearities. Whether market structures are monopolistically competitive depends in part upon *CR4*, in part upon the product differentiation variables discussed previously, and perhaps in part upon another Yale survey variable, *NICHES*, which assesses the extent to which company technological activities were aimed at designing products for specific market segments. Niche-targeting, as the discussion surrounding Figure 2.9 in Chapter 2 revealed, is a hallmark of monopolistic competition.

The progressive reduction of traditional tariff barriers under various GATT rounds did much to open the United States and other national markets to increased intra-industry trade during the 1960s and 1970s. However, other government actions worked in the opposite direction. With increasing frequency during the 1970s and 1980s, the U.S. government offered certain domestic industries special protection from imports through trade-barring actions. We take this phenomenon into account with the variable *BARS*, which is the 1971–1987 average of the sum of two zero-one dummy variables, one with unit value if an industry had trade barriers in place in any given year as a result of Section 201 "escape clause" actions, and the other with unit value in the first three years following the imposition of trade restraints under other sections of the applicable U.S. Trade Acts.

Another Yale survey variable, *INDIVIDUAL*, characterizes barriers to trade in a different way. It measures the extent to which company R&D activities were targeted toward customizing products to the needs of individual buyers. The more individualized product offerings are, the closer contacts must be between the manufacturer and its customers, and so the less likely trade over long distances and national borders will be.

Finally, we define two variables, *FDI:IN* and *FDI:OUT*, which measure, respectively, the percent of total 1981 payrolls in U.S. industries originating from plants owned by foreign parents, and 1982 payrolls of U.S.-based corporations' overseas subsidiaries as a percent of total relevant industry payrolls (of both multinational and other companies) in the United States.[14] Higher levels of intra-industry trade are hypothesized to

---

14. The main source of the inbound FDI ratios was U.S. Bureau of the Census, *Selected Characteristics of Foreign-Owned U.S. Firms, 1981* (Washington, D.C.: U.S. GPO, March 1983). The outbound FDI ratios came from U.S. Department of Commerce, Bureau of Economic Analysis, *U.S. Direct Investment Abroad: 1982 Benchmark Survey Data* (Washington, D.C.: U.S. GPO, December 1985), Table III.F7.

accompany greater direct foreign investment, especially when multinational firms specialize their production by nation and cross-ship diversified products to other markets. These are the least well measured of our variables, because relevant data were available only at levels of industry detail aggregated to 64 categories for inbound foreign direct investment positions and 25 categories for outbound investment. The aggregates had to be linked to 272 four-digit industries, either by entering the broader category's value for each narrowly defined industry or by aggregating the narrowly defined four-digit industries up to the same level of detail as the *FDI* variables.

Table 4.2 lists the explanatory variable mnemonics and provides a capsule description, along with mean values and standard deviations, for the 272 industry subsample. The Yale survey variables were measured along a seven-point Likert scale, with 7 meaning "very rapid" (for technological progress), "very relevant," or "very important," and 1 marking out

Table 4.2 Variables used to explain intra-industry trade

Variable	Description	Mean	Standard deviation
PROGRESS	Rate of product technology advance	4.52	1.10
SCIENCE	Relevance of basic science	3.81	.67
ENGINEERING	Relevance of academic engineering research	3.50	.78
STANDARD	Stress on product standardization R&D	4.02	.95
RD/S	1976–1977 R&D/sales (percent)	1.71	1.52
ADV/S	1977 advertising/sales (percent)	1.32	1.99
LEARNING	Importance of learning by doing on new products	5.10	.76
CR4	Four-firm concentration ratio (percent)	41.7	21.8
CR4SQ	CR4 squared	2,209	2,112
NICHES	Stress on niche-filling R&D	5.07	.87
INDIVIDUAL	Stress on individualizing products to customer needs	4.76	1.06
BARS	Import barriers, 1971–1987	.08	.19
FDI:IN	Percentage of U.S. payroll in foreign-owned plants	7.64	6.62
FDI:OUT	U.S. firms' overseas payrolls as percentage of domestic payroll	17.40	14.76

the opposite extreme. All other variables but *BARS* are scaled in per-centage terms.

Regression Analysis Results

Table 4.3 summarizes the results of various multiple regression analyses attempting to determine how intra-industry trade was influenced by the diverse explanatory variables. The analytic technique is ordinary least squares; *t*-ratios of the regression coefficients are printed in subscripted parentheses.

For a broad overview, regressions (3.1) and (3.2) average the *SH* index of intra-industry trade over the five years 1965, 1970, 1975, 1980, and 1985, first with all 449 four-digit industry observations and then for the higher-technology 272 industry subset on which variable matching is more exact. The differences between the two data sets are small, so we emphasize the better-measured subset. Several of our prior hypotheses are strongly supported. Intra-industry trade was greater, the more R&D-intensive an industry was, the more important learning by doing was, the more relevant academic engineering research (but not basic scientific research) was, and the less individualized products were to specific customers' demands. The influence of R&D intensity was stronger for the full sample than for the 272 industry subsample because the former (with an R&D/sales mean of 1.20 percent) contains more industries in which R&D was unimportant *and* intra-industry trade was correspondingly modest. Intra-industry trade was most extensive in middling oligopolies, with maximum *IIT* values occurring at four-firm concentration ratios of 46 percent in regression (3.1) and 50 percent in regression (3.2). A rapid rate of new product introduction during the 1970s (*PROGRESS*) does not appear to have affected intra-industry trade significantly, nor does advertising intensity capture the dimensions of product differentiation that lead to two-way trade. The results for the niche-filling variable are contrary to hypothesis for the full sample, as is the sign of the product standardization variable for both the full and the partial samples. There is a suggestion that intra-industry trade was greater in industries with high levels of specially imposed trade barriers. However, a breakdown of the data by years revealed that this effect was statistically significant in only one of the years (1985) actually covered by the 1971–1987 span of the *BARS* variable.

Regressions (3.3) and (3.4) repeat the same analysis substituting the Grubel-Lloyd index of intra-industry trade for our new index. The results

Table 4.3 Regressions explaining cross-sectional differences in intra-industry trade[a]

Explanatory variable	Regression number and dependent variable										
	(3.1) Average IIT_{SH}	(3.2) Average IIT_{SH}	(3.3) Average IIT_{GL}	(3.4) Average IIT_{GL}	(3.5) Average $LogIIT_{SH}$	(3.6) 1965 IIT_{SH}	(3.7) 1970 IIT_{SH}	(3.8) 1975 IIT_{SH}	(3.9) 1980 IIT_{SH}	(3.10) 1985 IIT_{SH}	(3.11) All years IIT_{SH}
RD/S	.215 (7.62)	.155 (4.04)	.0004 (0.19)	−.0068 (0.70)	.058 (3.27)	.168 (5.20)	.067 (1.81)	.106 (2.02)	.117 (2.25)	.296 (4.90)	.147 (6.87)
ADV/S	−.026 (1.45)	−.024 (0.83)	−.003 (0.59)	−.006 (0.81)	−.007 (0.50)	—	—	—	—	—	—
PROGRESS	.020 (0.43)	−.060 (0.96)	.036 (2.02)	.027 (2.81)	.031 (1.07)	−.047 (0.87)	−.147 (2.38)	−.092 (1.05)	−.064 (0.74)	−.011 (0.11)	−.060 (1.71)
SCIENCE	−.001 (0.01)	.010 (0.11)	.005 (0.25)	.040 (1.64)	−.001 (0.03)	—	—	—	—	—	—
ENGINEERING	.117 (2.68)	.225 (2.80)	−.007 (0.48)	.004 (0.21)	.077 (2.08)	.185 (2.82)	.315 (4.18)	.343 (3.22)	.329 (3.12)	.029 (0.24)	.248 (5.73)
STANDARD	.054 (1.32)	.061 (1.04)	.029 (2.21)	.021 (1.41)	.034 (1.27)	.051 (1.02)	.030 (0.52)	.033 (0.40)	.140 (1.72)	.063 (0.67)	—

LEARNING	.256 (4.50)	.247 (3.25)	.036 (2.02)	.054 (2.81)	.127 (3.65)	.142 (2.15)	.222 (2.93)	.303 (2.82)	.311 (2.93)	.256 (2.08)	.253 (5.78)
CR4	.0107 (1.58)	.0142 (1.46)	.0004 (0.19)	.0013 (0.53)	.0148 (3.30)	-.0008 (0.09)	.0103 (1.06)	.0110 (0.80)	.0221 (1.63)	.0243 (1.54)	.0135 (2.40)
CR4SQ	-.00012 (1.62)	-.00014 (1.42)	-.00001 (0.32)	-.00002 (0.65)	-.00012 (2.67)	.00001 (0.16)	-.00008 (0.81)	-.00014 (0.99)	-.00022 (1.57)	-.00024 (1.47)	-.00013 (2.28)
NICHES	-.075 (1.20)	.048 (0.57)	-.033 (1.71)	-.030 (1.42)	-.027 (0.70)	.004 (0.06)	.153 (1.92)	.168 (1.49)	.095 (0.85)	-.118 (0.91)	.053 (1.15)
INDIVIDUAL	-.085 (2.18)	-.159 (2.59)	.0009 (0.08)	-.009 (0.58)	-.067 (2.38)	-.158 (3.22)	-.189 (3.35)	-.227 (2.84)	-.192 (2.43)	.001 (0.01)	-.153 (4.69)
BARS	.265 (1.59)	.298 (1.03)	.0002 (0.00)	.008 (0.11)	.318 (2.38)	—	—	—	—	—	—
Intercept	-.548 (1.26)	-.749 (1.25)	.145 (1.07)	.066 (0.44)	-1.20 (4.38)	-.030 (0.06)	-.827 (1.57)	-1.19 (1.61)	-1.63 (2.21)	-.283 (0.33)	[5 values]
N	449	272	449	272	272	272	272	272	272	272	1,360
R^2	.226	.185	.071	.085	.259	.175	.142	.120	.145	.132	.165

a. *T*-ratios are given in subscripted parentheses.

are much weaker, although the *PROGRESS* variable becomes statistically significant and *SCIENCE* approaches significance for the higher-technology industry subsample. There is particularly marked deterioration of the R&D effect, apparently because high R&D intensity contributes to high *levels* of trade more than it ensures a relatively even balance of trade.

Because the distribution of IIT_{SH} is skewed to the right, with a five-year average value of 1.33 for 272 industries and a maximum value of 8.45, we must check whether the results might have been affected unduly by a few extreme observations. In regression (3.5), the dependent variable is transformed into logarithms (to base 10) to make its distribution more closely approximate normality.[15] Relative to non-logarithmic regression (3.2), no important changes in coefficient signs and *t*-ratios materialize, except that the *BARS* effect becomes statistically significant. Both the linear and nonlinear seller concentration terms also achieve statistical significance at the 99 percent confidence level, despite the significance-eroding effects of multicollinearity. Maximum intra-industry trade levels are found when the four leading U.S. producers had a combined 60 percent share of domestic shipments.

The sensitivity of results was also tested for variations in cutoff values applied to the standard deviation denominator of the IIT_{SH} index. All of the IIT_{SH} regressions in Table 4.3 are based on cutoff values of 3.5. Simple correlations across 272 industries between the IIT_{SH} indices with standard deviation cutoffs of 3.5 and those with 2.0, 5.0, and 8.0 cutoff values were 0.965, 0.984, and 0.922, respectively. Regressions for 272 industries, and including all the variables of equation (3.2) in Table 4.3 except *ADV/S*, *SCIENCE*, and *BARS*, were run for five-year average values of IIT_{SH}, assuming those four alternative cutoff values. There were no coefficient sign changes and only one lapse from statistical significance (on *INDIVIDUAL*, with $t = 1.92$ for a cutoff of 8.0). If anything, the "fit" was closer for alternative cutoff assumptions. Coefficient *t*-ratios were for no explanatory variable higher with cutoffs of 3.5 than with one of the three other cutoff values. R^2 values were 0.153, 0.178, 0.190, and 0.201 for cutoff values of 2.0, 3.5, 5.0, and 8.0, respectively. Thus, the choice of a 3.5 cutoff value had no distorting impact on the results.

---

15. Five zero or near-zero observations were plugged at 0.01, with a logarithmic value of -2. The mean of the transformed data was -0.005, the median 0.057, and the standard deviation 0.428.

Regressions (3.6) through (3.10) proceed year by year at five-year intervals with the *ADV/S*, *SCIENCE*, and *BARS* variables omitted, as in the standard deviation sensitivity analysis. The R&D, learning-by-doing, engineering relevance (excepting 1985), and product individualization variables perform robustly. Except in 1965, the market structure relationship has an inverted U-shape, with peak intra-industry trade levels occurring at *CR4* values in the range of 39 to 64. The niche-filling orientation variable has the sign predicted by theory in every year but 1985, but is statistically significant only for 1970.

Our measures of inbound and outbound foreign direct investment positions cover the years 1981 and 1982 only and are much more aggregated than the other data used in this chapter. The *FDI* variables were therefore used only in regressions (not separately reported in Table 4.3) with 1980 and 1985 intra-industry trade as the dependent variable. Tests were run for 272 industries, with relevant *FDI* values repeated in each matched narrowly defined industry, and with the data for 272 industries aggregated (using sales weights) to either 64 or 25 broader categories. Contrary to expectation, the *FDI* coefficient signs were erratic, but were more frequently negative than positive. None was statistically significant. Thus, no systematic pattern was found between foreign direct investment positions and intra-industry trade.

Finally, regression (3.11) pools the 272 industry subsample observations for all five years, letting each year have its own intercept value (rising significantly over time). Again, the results are consistent with patterns observed for the individual years, although *t*-ratios are generally higher because there are many more observations. The niche-filling variable has the sign predicted by theory and is statistically significant at the 87 percent confidence level. The collinear *CR4* and squared *CR4* coefficients are again significant, implying maximum intra-industry trade with domestic seller concentration levels of 51 percent. A test of whether the regression coefficients were homogeneous over time is accepted, with $F(32,1320) = 1.10$. Thus, even though some individual variables perform erratically, there was significant consistency over the 1965–1985 period in the broad pattern of intra-industry trade relationships.

Conclusion

The concept of intra-industry trade has played a prominent role in theories of comparative advantage and trade structures for nations like the

United States. Using a new measurement technique, we have found that intra-industry trade involving U.S. manufacturing industries increased appreciably between 1965 and 1985. There were distinct patterns consistent with some of the principal theoretical hypotheses. Intra-industry trade was greater in industries with an orientation toward developing new products, manifested in high ratios of research and development spending to sales, substantial perceived relevance of academic engineering research, and (less strongly) a strategic emphasis on filling niches in product characteristics space. Intra-industry trade thrived where substantial product-specific economies of scale could be achieved through learning by doing. The influences of plant-level scale economies and open-entry oligopolistic competition are suggested by an inverse U–shaped relationship between intra-industry trade and domestic market concentration. Intra-industry trade was significantly lower when companies geared their product development efforts to meeting the specialized needs of individual customers. No systematic influence from foreign direct investment positions was discernible.

Appendix: Compiling the Data on Imports and Exports

Data on U.S. exports and imports are collected under classification schemes that differ in many ways from the Standard Industrial Classification (SIC), which forms the backbone of statistics on industry output, research and development, profitability, and much else.[16] To transform the import and export data to an SIC basis, concordances have been devised by the U.S. Department of Commerce and private scholars. Complicating matters is the fact that the SIC for manufacturing industries was substantially revised in 1972 and again in 1987, as were the import and export classifications at various times. As a result, even though trade

---

16. Throughout the period covered by our series, imports were classified under the Tariff Schedule of the United States and exports under a "Schedule B" classification. In 1989, a new "Harmonized System" was adopted for both imports and exports. On the cross-classification problems, see U.S. Bureau of the Census, *U.S. Foreign Trade Statistics: Classifications and Cross-Classifications, 1980* (Washington, D.C.: U.S. GPO, February 1981); and William B. Sullivan, "The 1987 Standard Industrial Classification (SIC), the Harmonized System (HS), and Cross Classification HS to SIC," draft report, U.S. Department of Commerce, Office of Trade Administration, January 12, 1990.

statistics reorganized to an SIC basis have been published regularly, the individual reports are not fully comparable over time.[17]

Our task was greatly alleviated by the availability of two SIC-based import and export compilations, one (covering 1972–1987) from the U.S. Department of Commerce Office of Trade Administration[18] and the other (spanning 1963 through 1985) from a National Bureau of Economic Research (NBER) effort.[19] The two series accepted quite different assumptions in bridging the U.S. Tariff Schedule and SIC gaps. The Office of Trade Administration series allocated each block of product trade data to the single SIC category where it fitted best, accepting two kinds of errors—overassignment of values to SICs in which the Tariff Schedule categories were broader than the SIC categories, and concomitant underassignment of values to the SIC categories where the fit was inferior. As a result, many SIC industries are shown not to have had any imports at all, even though they did in fact experience inbound trade. The National Bureau of Economic Research approach was to divide excessively broadly defined trade flows among appropriate SIC industries, using historical patterns and data on the relative size of the industries as a basis.

Neither source perfectly satisfied our need for a complete time series covering all of the 449 four-digit manufacturing industries for which other industry-level data were defined.[20] The more comprehensive NBER compendium was used as a starting matrix, modified through insertion or substitution of data from the Office of Trade Administration and primary sources where appropriate. A number of errors were found and corrected by comparing the two series, industry by industry, and checking discrep-

---

17. The principal source was U.S. Bureau of the Census, *U.S. Commodity Exports and Imports as Related to Output,* various years. Budgetary stringencies led to the report's termination after 1982. Imports were reported on a consistent c.i.f. basis (i.e., delivered to a U.S. entry port, with insurance and freight paid) beginning only in 1974. A splice to earlier years' data, reported on a port of embarcation basis, is included in the report for 1974.

18. An incomplete version was published in April 1988 by the Industry Statistics Division, International Trade Administration, U.S. Department of Commerce, under the title *U.S. Trade Data, 1974–1987.* I am indebted to William Sullivan of that office for making the complete data available in machine-readable form and for explaining intricacies of the concordance problem.

19. The detailed linking was done by John Abowd. Larry Katz kindly provided the data in machine-readable form.

20. The number of four-digit manufacturing industries also varies over time, but we standardized on 449, which reflected the composition of the Standard Industrial Classification as of 1982.

ancies against primary data sources. For nine industries, OTA import data were substituted into the NBER series. Thirty reverse substitutions from NBER to OTA were required to fill gaps in OTA coverage. For four industries whose coverage by primary trade statistics and hence by both the OTA and the NBER compilations was particularly deficient, average import ratios for the entire manufacturing sector were inserted. The modified NBER series was extended from its 1985 terminus by chain-linking OTA figures for the years 1986 and 1987.

For the years 1972–1985, on which the two series overlapped, the ratios of imports and exports to domestic output value from the OTA and NBER series, reconstructed by us, were highly correlated, with simple (Pearsonian) correlation coefficients of 0.892 for imports and 0.903 for exports. The NBER-based series are better suited to our needs in Chapters 4 and 5, so the analyses there use them. They are imperfect, but we consider them sufficiently reliable for our statistical purposes.

5 R&D Reactions to Import Competition

If comparative advantage comes in part from having superior products and production processes, the structure of U.S. trade flows must depend upon how domestic manufacturers react to technological challenges from abroad. Advantage in exporting high-technology merchandise is more likely to be sustained when domestic firms increase their own R&D in response to intensified innovative efforts by foreign rivals—that is, when they react aggressively—than when they cut back their R&D (a "submissive" reaction). In this chapter we seek quantitative insight into the pattern of U.S. firms' reactions.

Two Pilot Studies

The analysis of innovative rivalry between domestic and foreign enterprises is constrained by the availability of data. Two pilot studies using data in the public domain provided preliminary insights.[1] We review them briefly and then report a more ambitious analysis using micro-data from the U.S. Census Bureau.

As the case studies in Chapter 3 reveal, Japanese companies have marshalled the most far-reaching technological challenges to U.S. companies. We therefore ask whether there are detectable interactions between Japanese and American industrial R&D spending. Data were available only at the all-industry level. They were drawn from the 1967–1982 series

---

1. The pilot studies appear in F. M. Scherer, "International R&D Races: Theory and Evidence," in Lars-Gunnar Mattsson and B. Stymme, eds., *Corporate and Industry Strategies for Europe* (Amsterdam: Elsevier, 1991), pp. 117–137.

compiled by David Levy and Nestor Terleckyj,[2] with three additional years linked from U.S. National Science Board compilations.[3] Inflation was removed through deflation by national price indices. The Japanese data were converted to U.S. dollars using a constant 1975 purchasing power parity ratio, avoiding the problem of variation introduced by fluctuating exchange rates. The U.S. data cover company-financed R&D only; the Japanese series includes all industrial R&D, of which only a small but imperfectly estimable fraction was financed by the government.

A standard Granger-Sims test yielded evidence of two-way simultaneous influence; that is, U.S. and Japanese R&D outlays tended to rise and fall concurrently, perhaps because of mutually stimulating rivalry but at least as plausibly owing to some common external influence such as changes in technology or macroeconomic conditions. A test for changes in the reaction pattern over time revealed that U.S. R&D outlays *fell* with increased Japanese R&D expenditures during the late 1960s, but the changes became increasingly positive and parallel by the late 1970s and early 1980s. The data were at too aggregate a level to illuminate chains of causation confidently.

The second pilot study examined the relationship between the penetration of U.S. markets by imported products and the lagged R&D spending of 26 companies specialized in seven industries—steel, machine tools, anti-friction bearings, consumer radio and television sets, semiconductors, automobiles, and photographic supplies and equipment. Data were available only for the years 1975–1985. The panel's starting point was dictated by a Financial Accounting Standards Board requirement that U.S. corporations publicly disclose "material" R&D expenditures beginning in 1975. The merger wave of the 1980s wrought such widespread transformations of corporate structures that after 1985, it became difficult to sustain meaningful R&D time series for representative companies.

Statistical analysis revealed that on average, companies tended at first to reduce their R&D/sales ratios in response to increasing import inroads—that is, to react *submissively* to intensified foreign competition. But in the later years of the 1975–1985 sample period, their reactions

---

2. David M. Levy and Nestor E. Terleckyj, "Trends in Industrial R&D Activities in the United States, Europe and Japan, 1963–83," National Planning Association paper presented at a National Bureau of Economic Research conference, August 1985.

3. U.S. National Science Board, *Science and Engineering Indicators: 1987* (Washington, D.C.: U.S. GPO, 1988), pp. 293–294.

became increasingly aggressive. Further investigation disclosed that observed reaction patterns varied appreciably across the seven industry groups. In steel and consumer electronics, reactions were on average much more submissive than for the sample as a whole; in semiconductors and photo equipment, the reactions were palpably more aggressive. Thus, consistent with the case study evidence, no single homogeneous behavioral pattern stood out. The statistical evidence on consumer electronics and photo equipment parallels the qualitative case histories on those industries.

Drawing a Broader Statistical Sample

The pilot study results were sufficiently interesting that a major sample expansion effort appeared worthwhile. By far the most comprehensive and reliable source of data on U.S. companies' R&D expenditures, defined in a standardized and consistent way, is the survey conducted annually since 1957 by the U.S. Census Bureau on behalf of the National Science Foundation.[4] The survey results are published at levels of aggregation too broad to be of use in our inquiry. To obtain access to individual company data, the author worked as a research fellow at the U.S. Census Bureau between 1990 and 1991.

Our analysis here utilizes Census data for 308 companies covering the years 1971 through 1987. Several factors constrained the sample coverage in space and time.

Our goal is to determine how the R&D spending of U.S. companies reacts to technological challenges from abroad, manifested primarily in imports of high-technology goods into U.S. markets. Because approximately 97 percent of all industrial research and development in U.S. industry is performed by manufacturing firms, the sample was limited to companies whose primary activity was manufacturing. Many industries have only a weak R&D orientation. Member companies perform little R&D on their own and rely upon the products and services of other industries for most of the new technology they implement.[5] Although it

---

4. The first survey was published in the National Science Foundation report, *Funds for Research and Development in Industry: 1957* (Washington, D.C.: U.S. GPO, August 1960). Subsequent annual reports are typically titled *Research and Development in American Industry*.

5. See F. M. Scherer, "Interindustry Technology Flows in the United States," *Research Policy*, 11 (August 1982), pp. 227–245.

is difficult to draw a precise line, an attempt was made to exclude such cases by omitting companies whose primary industry affiliation was in Standard Industrial Classification groups 20–25 (food, tobacco, textiles, clothing, wood products, and furniture), 27 (publishing), and 31 (leather goods). In addition, companies at home in the more technology-intensive petroleum refining industry group (SIC 29) were excluded because import patterns were distorted during the sample period by two OPEC shocks and extensive government regulation of prices and product allocations.

The Census Bureau forms a new sampling panel for its annual R&D surveys at intervals of roughly five years—for the period covered by our sample, in 1971, 1976, 1981, and 1987. In transition years, more than 10,000 companies are surveyed. Those whose R&D outlays fall below specified threshold values, ascending from $200,000 in 1971 to $1 million in 1981, are omitted from surveys for the next four years, but their names are retained in the NSF survey files, and R&D spending values for them are "imputed" on a formula basis. All such intermittently surveyed companies were excluded from our sample.

Companies were dropped from the sample used here for several further reasons. Like other government surveys, the R&D survey suffers from non-response problems. Census staff routinely impute R&D spending estimates for non-responding companies. An attempt was made to identify such imputations and to exclude companies on which imputations were made for two or more years in a row. The remaining single-year imputations were recalculated by averaging the R&D/sales values for the previous and subsequent years. In the course of this effort, several dozen implausible R&D values were discovered. Each case was checked against original R&D survey forms when available (among other things, to identify coding errors) and against other Census and published historical records (to identify changes in company structure and the like). Corrections were made when possible; otherwise the companies were deleted from the sample.[6] A few companies that would otherwise have qualified for sample inclusion were dropped because complementary Census records were incomplete.

The time series analysis starting date was set at 1971 because data for

---

6. For example, a few companies reported sharp R&D jumps in the early 1980s, possibly to take advantage of the 1981 R&D tax credit law by redefining what was "R&D." They were excluded from the sample.

earlier years proved to be inaccessible.[7] The terminal date was set at 1987 because major changes in the Standard Industrial Classification precluded reliable links of the R&D data to other Census statistics for subsequent years. The company panels surveyed continuously for five years typically contained more than 1,000 companies. However, there was considerable turnover from one panel to another as a result of mergers, sell-offs, growth of companies to "certainty" sampling size, and other structural changes. The requirement that 17 years of continuous data be available plus the elimination of petroleum and low-technology industry specialists left a basic working sample of 362 companies.[8] Data quality controls led to the deletion of 54 companies from that group, leaving the final sample count at 308.

Those 308 companies accounted for 62 percent of all company-financed industrial research and development in the United States for 1972, 65 percent in 1974 (the year of peak coverage), 61 percent in 1980, and 49 percent in terminal year 1987. The sample's declining share of the total R&D universe suggests an important omission: rapidly growing high-technology companies too small to qualify for inclusion in the 1971 and 1976 Census continuous survey panels. On average across all years, company-financed R&D outlays were 3.22 percent of domestic sales, rising from a minimum average of 2.83 percent in 1974 to a maximum of 4.08 percent in 1986. The 308 sample companies conducted 56 percent of federal government–funded industrial R&D in 1972 and 51 percent of such R&D in 1987.

Controlling for Company Structure Changes

Mergers, sell-offs, and other corporate restructurings pose an analytic problem transcending their impact on sample selection. With one short-lived exception, reliable industrial R&D data are available only at the whole-company level. But consider what happens when a company such as ITT, originally specialized in the high-technology telecommunications

---

7. A computer tape believed to have contained R&D survey data for 1968–1971 was found, but it was coded in an archaic Univac binary language whose translation would have retarded analytic progress unacceptably.

8. In a few cases, large companies acquired by another enterprise in 1986 or 1987 were retained in the sample even though unconsolidated R&D data for the post-acquisition years were unavailable, having been amalgamated with the new parent's report. The total number of company-year observations lost in this way was 26.

business, acquires a sizable low-technology company such as Continental Baking. The company's R&D/sales ratio drops abruptly from one year to the next for no reason plausibly connected with import competition. The opposite happens when, as in 1984, ITT divested Continental Baking.[9]

To control for such structural changes, which were of great importance due to a major sell-off wave during the 1970s and a merger wave in the 1980s, we adopt a novel technique. We define a variable:

$$[5.1] \qquad RDINDEX_{it} = \sum_{j=1}^{449} w_{ijt}(RD/S)_{j,77},$$

where $(RD/S)_{j,77}$ is the average 1976–1977 ratio of R&D to sales in the jth industry occupied by company i,[10] and w_{ijt} is the share of company i's total domestic manufacturing establishment sales contributed by four-digit industry j during year t.[11] Thus, $RDINDEX$ is an industry-weighted average telling what company i's R&D/sales ratio *would be* if the company pursued R&D exactly as intensively in each of its domestic lines as all surveyed firms in those lines did during 1976 and 1977. Changes in company structure, most prominently through mergers and sell-offs, lead to changes in weighting variable w_{ijt} and hence in $RDINDEX$. The average value of $RDINDEX$ across 308 companies drifted upward over time from a minimum value of 2.31 percent in 1973 to 2.51 percent in 1987, indicat-

---

9. Even more severe problems would intrude if the analysis focused on R&D *levels* rather than R&D/sales ratios, because every sizable merger, and not just those among firms of differing R&D intensities, introduces a time series discontinuity. We analyze R&D/sales ratios rather than levels because the technique described below allows us to control for structural changes affecting ratios.

10. The R&D/sales ratios are averaged from Federal Trade Commission Line of Business surveys for the years 1976 and 1977. See Federal Trade Commission, Bureau of Economics, *Statistical Report: Annual Line of Business Report, 1976 and 1977* (Washington, D.C., May 1982 and April 1985). The last such survey, providing statistics on industry R&D expenditures disaggregated to 220 industry categories, was for 1977. When a figure for one of the two years was unavailable, the ratio for the other year was used alone. When an FTC "line of business" category was more broadly defined than a four-digit Standard Industrial Classification industry, the broader group's R&D/sales ratio was used in each four-digit industry.

11. The sales (technically, value of shipments) data are drawn from Census of Manufactures and Annual Survey of Manufactures establishment (plant) records combined under the Census Bureau's Longitudinal Research Database (LRD) program. See Robert H. McGuckin and George A. Pascoe, "The Longitudinal Research Database (LRD): Status and Research Possibilities," U.S. Bureau of the Census, Center for Economic Studies, Discussion Paper 88-2 (July 1988).

ing that the sample companies' sales mix gradually moved toward more R&D-intensive industries.

We use as the basis of our principal R&D variable in the analyses that follow an adjusted R&D variable:

[5.2] $ADJRD_{it} = (RD/S)_{it} - RDINDEX_{it}$,

where $(RD/S)_{it}$ is the ratio of company i's self-financed R&D conducted in the United States to its domestic sales. The R&D variables are henceforth scaled uniformly in percentage terms, that is, as ratios $\times$ 100.

RDINDEX in effect controls for structural differences in what students of R&D have called "technological opportunity."[12] Table 5.1 tests its effectiveness in doing so. In regression (1.1), annual observations on R&D as a percentage of sales for the 308 sample companies are regressed on *RDINDEX* as the sole explanatory variable. The percentage of variance explained (r^2) is 0.492, surpassing the explanatory power achieved in earlier studies using fixed industry effects or survey-based variables to measure technological opportunity.

Regression (1.2) adds one-zero dummy variables for each year 1972–1987 and a price-cost margin variable PCM_{it}, which provides an index of firms' gross profitability.[13] The dummy variable coefficients exhibit a cyclical pattern, with R&D outlays, which tend to be relatively sticky from one year to the next, falling less than sales in recession years 1975 and 1982. A rising trend in the late 1970s and mid-1980s is also evident. The coefficient of 1.247 on *RDINDEX* after correcting for year effects shows that our sample companies were over-achievers relative to the averages for the industries in which they operated. This selection bias occurred because companies had to exceed certain R&D spending thresholds consistently to remain in the Census Bureau's continuous survey panels. The *PCM* coefficient is significantly positive, confirming either of

---

12. See F. M. Scherer, "Firm Size, Market Structure, Opportunity, and the Output of Patented Inventions," *American Economic Review*, 55 (December 1965), pp. 1097–1123; W. L. Baldwin and John T. Scott, *Market Structure and Technological Change* (Chichester: Harwood, 1987), pp. 105–109; and Wesley M. Cohen and Richard C. Levin, "Empirical Studies of Innovation and Market Structure," in Richard Schmalensee and Robert D. Willig, eds., *Handbook of Industrial Organization*, vol. 2 (Amsterdam: North-Holland, 1989), pp. 1083–1090.

13. *PCM* = (value of shipments − materials costs − payroll costs − supplementary benefit costs)/(value of shipments) for company i's domestic manufacturing plants in year t. It is scaled in ratio form. Costs and outlays not deducted include depreciation, selling and R&D costs, other corporate overhead, and income tax liabilities.

Table 5.1   Regression analysis of annual R&D/sales levels[a]

Explanatory variable	Regression number			
	(1.1)	(1.2)	(1.3)	(1.4)
Intercept	.091 (1.68)	−.70 (5.01)	−1.01 (6.08)	−.90 (5.45)
RDINDEX	1.323 (71.01)	1.247 (65.91)	1.289 (57.48)	1.289 (57.59)
PCM	—	2.931 (12.95)	2.537 (9.49)	2.515 (9.42)
DUM72	—	−.13 (0.76)	s[b]	s
DUM73	—	−.24 (1.37)	s	s
DUM74	—	−.29 (1.67)	s	s
DUM75	—	−.10 (0.55)	s	s
DUM76	—	−.15 (0.84)	s	s
DUM77	—	−.08 (0.47)	s	s
DUM78	—	−.17 (0.97)	s	s
DUM79	—	−.13 (0.76)	s	s
DUM80	—	.08 (0.49)	s	s
DUM81	—	.20 (1.17)	s	s
DUM82	—	.63 (3.59)	s	s
DUM83	—	.63 (3.63)	s	s
DUM84	—	.53 (3.04)	s	s

Table 5.1 *(continued)*

Explanatory variable	Regression number			
	(1.1)	(1.2)	(1.3)	(1.4)
DUM85	—	.66 (3.80)	s	s
DUM86	—	.78 (4.44)	s	s
DUM87	—	.69 (3.88)	s	s
TECH	—	—	.302 (3.84)	—
MBA	—	—	− .034 (0.47)	—
LAW	—	—	.246 (2.82)	—
UNCT	—	—	− .006 (0.04)	−
ONLYTECH	—	—	—	.138 (1.49)
ONLYMBA	—	—	—	− .181 (1.22)
ONLYLAW	—	—	—	− .040 (0.29)
TECH + MBA	—	—	—	.117 (1.13)
TECH + LAW	—	—	—	.757 (5.90)
MBA + LAW	—	—	—	− .122 (0.76)
R^2	.492	.520	.563	.566
N	5,210	5,210	3,718	3,718

a. *T*-ratios are given in subscripted parentheses. The intercept in regressions (1.2)–(1.4) is the 1971 deviation from the overall mean.

b. s: Coefficient suppressed.

two plausible hypotheses: that larger margins induce higher R&D spending under some variant of the Dorfman-Steiner theorem, or the necessity in the long run for quasi-rents to be elevated enough to cover R&D expenditures.[14]

Regressions (1.3) and (1.4) introduce an additional set of variables to help explain the sample companies' observed R&D/sales percentages. Through a considerable research effort, it was possible to identify the educational background of the top two executive officers (usually, the chairman and president, or in ambiguous cases, the chief executive and operating officers) of 221 sample companies.[15] Companies were coded as to whether those leaders included a person with university-level training in science or engineering (*TECH*), law (*LAW*), and/or graduate business administration (*MBA*). For our sample of relatively technology-intensive companies, the proportion of firms with a technically educated leader rose over time to a peak of 71.5 percent in 1980 and declined by five percentage points thereafter. On average, 30.7 percent of the companies had MBA holders in at least one of their top positions. The presence of MBAs rose at an increasing rate. For legally trained executives, the trend was downward, with an average of 18.7 percent over the 17-year period.

Regression (1.3) adds zero-one dummy variables indicating whether a company had a top executive in one of these three categories plus a dummy variable, *UNCT*, for 239 one-year observations on which it was not clear whether a company leader had technical training. The statisti-

---

14. On the Dorfman-Steiner theorem's predictions, see Douglas Needham, "Market Structure and Firms' R&D Behavior," *Journal of Industrial Economics*, 23 (June 1975), pp. 241–255. The coefficient value implies more support for the short-run Dorfman-Steiner view, since a ten percentage point increase in *PCM* is associated on average with a 0.29 percentage point increase in the R&D/sales variable, ceteris paribus.

15. Company executives were identified through annual listings in Standard & Poor's *Register of Corporations, Directors and Executives*. Educational backgrounds could usually be discerned when executives' biographies appeared in periodic issues of *Who's Who in Finance and Industry*. When a single executive held both the chief executive officer and chief operating officer titles, or when for other reasons only a single leader was identified, only the background of that executive was coded.

Years during which the principal U.S. companies appearing in the Chapter 2 case studies had technically educated executives were as follows (with the "U" designation identifying uncertain cases): Boeing, 1971–1987; Caterpillar, 1972–1984, U1985–1987; Corning, 1971–1987; Eastman Kodak, 1971–1987; Firestone, 1977–1978, 1983–1987, U1979; General Electric, 1972–1987; Gillette (none); Goodrich, 1971–1978, U1987; Goodyear, 1971–1977, 1983–1987, U1978–1982; Hewlett-Packard, 1971–1987; ITT, 1971–1987; McDonnell-Douglas, 1971–1987; RCA, 1971–1975; Texas Instruments, 1971–1987; Uniroyal, 1971–1987; Western Electric, 1971–1987; Xerox, 1977–1987; Zenith, 1971–1982.

cally significant coefficient for *TECH* reveals that, controlling also for profitability and the R&D intensity of industries occupied by the companies, having a top executive with a scientific or engineering background was associated with R&D/sales percentages higher by 0.30 points.[16] (For the 3,718 observation sample, R&D outlays averaged 3.12 percent of sales.) R&D intensity was higher by an almost equivalent amount when a legally trained officer was present. The effects for MBAs and uncertain cases were small and statistically insignificant.

In many companies, leadership was shared between a lawyer and a technologist, and in other cases, a single executive had undergraduate training in science or engineering plus graduate work in law or business. Regression (1.4) categorizes the data more richly, as follows:

ONLYTECH   Company had technically trained leader but no law-
                 yer or MBA.
ONLYLAW   Company had a legally trained leader, but no tech-
                 nologist or MBA.
ONLYMBA   Company had an MBA leader, but no lawyer or
                 technologist.
TECH + MBA   Leaders had *both* technical and graduate business
                 backgrounds.
TECH + LAW   Leaders had *both* technical and legal backgrounds.
MBA + LAW   Leaders had both MBA and law backgrounds, but
                 no technical education.

The base case consists, as in regression (1.3), of companies whose leaders had only non-technical bachelors' degrees or no known degrees. Regression (1.4) suggests the combination of technical and legal education to have been especially conducive to the support of R&D, adding 0.76 points on average to R&D as a percentage of sales. Technical backgrounds without augmentation from other advanced degrees (comprising 39.6 percent of all cases) added 0.14 points—a coefficient significantly different from zero only at the 93 percent confidence level. Having officers with legal or MBA training alone or a combination of legal and MBA training (6.7, 8.2, and 4.9 percent of the cases, respectively) was *negatively* asso-

---

16. Because the sample composition changed, the R^2 values in regressions (1.2) and (1.3) are not comparable. The R^2 value for the analogue of regression (1.2) with the smaller (221 company, 3,718 observation) sample was 0.561.

ciated with R&D/sales ratios, although none of the three effects approaches statistical significance.

These results cannot prove a causal relationship running from technical education to more vigorous support of R&D. It might be argued, for example, that the technical background coefficients reflect a company size influence rather than a research-intensive orientation stemming from education. *TECH* was positively but weakly correlated with the logarithm of current-year company sales, with $r = 0.036$. A tenfold increase in company size was associated with a 1.3 percentage point increase in the likelihood of having an engineering- or science-trained leader. However, when $Log_{10}SALES$ was introduced as a separate explanatory variable (not reported in Table 5.1), the values of the technical background coefficients were altered only trivially. Other more subtle influences could have been at work, but the results at least suggest, subject to appropriate caveats, that educational background mattered.

Measuring Foreign Competition

We advance now toward our principal goal, which is to determine how U.S. companies' R&D spending was affected by the extensive growth of high-technology competition from foreign firms during the 1970s and 1980s. Data permitting an analysis of how U.S. firms' R&D responded to changes in overseas rivals' R&D are not available. We therefore measure the vigor of foreign competition by changes over time in two measures of import penetration into U.S. markets—the ratio of imports to the value of domestic plant shipments, and the ratio of net exports, that is, exports minus imports, to domestic shipment value. The net export variable is included because in some industries (such as automobiles and computers) multinational corporations engage in significant amounts of intra-industry, intra-company trade, simultaneously exporting and importing similar (but differentiated) products. In 1987, for example, exports of $65.2 billion, or approximately 26 percent of total U.S. merchandise exports, went to majority-owned overseas affiliates of U.S. multinational corporations. Imports of $55.9 billion, or 13.6 percent of total U.S. merchandise imports, flowed from offshore affiliates of U.S. multinationals to the U.S. parents' domestic units.[17] To the extent that such

---

17. See Raymond J. Mataloni, "U.S. Multinational Companies: Operations in 1988," *Survey of Current Business* (June 1990), p. 42; and U.S. Department of Commerce, Bureau

intra-company trade flows are evenly balanced, they cancel each other out in a net exports measure but not with an imports measure. The imports measure in such cases is likely to overstate the amount of import competition truly present.

Several of our case studies revealed that technological challengers from abroad established production operations in the United States in addition to, or (in fewer cases) instead of, exporting competing products to the U.S. market. Competition through foreign direct investment (FDI) became increasingly prominent during the 1980s.[18] Import data alone therefore underestimate the amount of competition confronting U.S.-based enterprises. If FDI magnitudes are positively correlated with import penetration, estimates of domestic firms' R&D reactions to foreign competition using import data alone will be exaggerated. If FDI replaces imports so much that the two are negatively correlated, reactions estimated using import data alone will be biased toward zero. Data on foreign firms' manufacturing activities in the United States are available only at higher levels of industry aggregation, and for less complete time series, than the import data used in the remainder of this chapter. For 64 (mostly) three-digit aggregations spanning the same SIC codes as our main sample, the correlations between the 1981 payrolls of foreign-owned plants as a percentage of total U.S. industry payrolls and industry imports as a fraction of total U.S. output ranged from $+0.16$ to $+0.19$, depending upon the time lag assumed.[19] The correlations tended to increase with proximity in time of the two indices; that is, import penetration ratios for 1981 and 1982 were slightly more correlated with 1981 FDI ratios than were import ratios for the early and mid-1970s. However, none of the correlations was statistically significant, which suggests that any positive bias imparted by using import penetration as a proxy for foreign competition more generally is modest.

---

of Economic Analysis, *U.S. Direct Investment Abroad: Revised 1987 Estimates* (Washington, D.C.: U.S. GPO, July 1990), Table 51. In 1982, the comparable import figure was 15.8 percent. U.S. Bureau of Economic Analysis, *U.S. Direct Investment Abroad: 1982 Benchmark Survey Data* (Washington, D.C.: U.S. GPO, December 1985), p. 335. Merchandise export estimates are taken from the *Economic Report of the President*.

18. See Robert E. Lipsey, "Foreign Direct Investment in the United States and U.S. Trade," National Bureau of Economic Research working paper no. 3623 (February 1991).

19. The foreign-owned plant payroll data are from U.S. Bureau of the Census, *Selected Characteristics of Foreign-Owned U.S. Firms* (Washington, D.C.: U.S. GPO, March 1983). Publication of the series was discontinued after 1982, so comparable data for the period when inbound FDI increased appreciably are not available.

The import and net export variables employed in the analysis that follows came from the sources and "cleansing" procedures described more fully in the appendix to Chapter 4. They are necessarily collected at the industry or product line level. They are linked to sample companies by computing weighted averages, for instance, of import/(value of shipments) ratios $(IMP/VS)_{jt}$ in industry j multiplied by w_{ijt}, the share of company i's sales recorded in industry j during year t. Like R&D/sales ratios, the weighted average import competition values can vary, sometimes wildly, with changes in company structure over time. We control for company structure changes by defining an import index variable, which is the sales-weighted share of imports (or net exports) as a percentage of domestic plants' value of shipments, averaged over the base years 1978–1980 (which lie at the midpoint of our time series). Our adjusted measures of import competition for company i are therefore:

[5.3] $$ADJIMP_{it} = \sum_{j=1}^{449} w_{ijt}(IMP/VS)_{jt} - \sum_{j=1}^{449} w_{ijt}(IMP/VS)_{j,78-80}$$

and

[5.4] $$ADJNX_{it} = \sum_{j=1}^{449} w_{ijt}(NX/VS)_{jt} - \sum_{j=1}^{449} w_{ijt}(NX/VS)_{j,78-80}$$

for imports and net exports, respectively. The scaling, again, is in percentage terms.

The simple average value of IMP/VS for all companies and years was 9.45 percent;[20] the average for NX/VS was 2.40 percent. The average value of the import index (the right-hand variable in equation [5.3]) was 8.62 percent; the average net export index value 3.61 percent. The index variables exhibited weak time trends, with the import index drifting downward $(r = -0.069)$ and the net export index variable upward $(r = 0.057)$ over the years 1971–1987. This implies a tendency for company sales in high import–impact industries to have declined relative to those in less affected industries as import penetration into U.S. markets rose.

Reactions to Import Competition: Time Series Analysis

Our basic hypothesis, derived from the theoretical analysis in Chapter 2 and the case studies in Chapter 3, is that U.S. firms alter their R&D/sales

---

20. The trend in annual averages of this variable is shown by the dashed line ("308 Companies") in Figure 4.1.

ratios in response to changes in technology-based import competition, exemplified by (mostly) rising import/output ratios. This implies a regression relating first differences in R&D/sales ratios to first differences in measures of import penetration:

[5.5] $\Delta ADJRD_{it} = a + b_1 L(\Delta ADJIMP)_{it} + b_2 X_{it} + e_{it}$,

where b_1 is a reaction coefficient, $L(\cdot)$ is a lag operator on the first differences of the index-adjusted import variable, and X is a matrix of business conditions and other relevant variables.

Because industry import and export data linked to the company level measure the intensity of import competition faced by specific firms only imperfectly, and because high-technology competition from foreign firms may materialize in foreign markets or through direct investment in the U.S. market, $\Delta ADJIMP$ measures with error the competitive challenge we seek to characterize. As we have seen, the positive correlation observed between import penetration and foreign direct investment could cause estimated reaction coefficients to be exaggerated. However, to the extent that errors in our key explanatory variables are largely random, the b_1 coefficients will be biased toward zero, especially in a first differences time series specification.[21]

The index technique used to control for changes in company structure created another statistical difficulty. Unusually large changes in $\Delta ADJRD$ and (especially) the adjusted import ratio first differences sometimes materialized, especially when multi-line plants of narrowly specialized companies experienced sales mix changes, causing the plant's industry classification to jump from one Standard Industrial Classification category to another. Graphic plotting revealed the first differences of *ADJRD*, *ADJIMP*, and *ADJNX* to have distributions that conformed imperfectly to the normal Gaussian pattern. The distributions were appropriately peaked near mean values, but they had long, thin tails on both sides.[22] Sensitivity tests indicated that the extreme values had little impact on estimated reaction coefficient signs and magnitudes, but because of the disproportionate weight they received in least squares regressions, they forced the coefficients' standard errors upward and masked the signifi-

---

21. See Zvi Griliches and Jerry A. Hausman, "Errors in Variables in Panel Data," *Journal of Econometrics*, 31 (February 1986), pp. 93–118.

22. In this respect they resembled the Cauchy distribution, which asymptotically has no finite variance. See Howard W. Alexander, *Elements of Mathematical Statistics* (New York: Wiley, 1961), pp. 117–119.

cance of relationships among less extreme observations. To deal with this outlier problem, common in detailed micro-data sets, we deleted 104 observations on which either $\Delta ADJRD$ or an adjusted import competition variable lay more than four standard deviations from its mean.[23] Sensitivity tests revealed coefficient sign patterns to be unaltered over alternative truncation thresholds, although the statistical significance of estimated coefficients declined as the thresholds were moved well above and well below four standard deviations.

Other Explanatory Variables

Several additional explanatory variables are used in the time series analysis of R&D/sales ratio changes.[24]

As we have seen in Table 5.1, there were both cyclical and trend patterns in company R&D/sales ratio movements over the 1971–1987 period. These are taken into account most powerfully by introducing dummy variables for each year, although in certain cases we shall instead employ an index of business conditions $BUSCON_{it}$, which measures year-to-year percentage changes in real manufacturing GNP. In constructing *BUSCON*, a distinction was made between non-durable and the more cyclically sensitive durable goods industries. The two components were weighted to the company level by the yearly shares of individual company sales in the durable and non-durable categories.

We found in Table 5.1 that R&D/sales levels were higher, the wider price-cost margins were. For our first differences analysis, we introduce

---

23. Inspection revealed that most of the deleted observations were the result of industry classification or measurement errors rather than meaningful structural changes. Five companies with extreme values in several years were deleted altogether from the sample, reducing the final sample to 308 companies. For the 308 company sample after deletion of extreme observations, the mean year-to-year change in the R&D/sales ratio was 0.033 percentage points, with a standard deviation of 0.75 points. Before deletion, the mean was 0.039 points and the standard deviation 0.87 points. For $\Delta ADJIMP$, the mean year-to-year change before deletion of extreme observations was 0.77 percentage points; after winnowing, it was 0.69 points. The standard deviations were 4.07 and 2.10 points, respectively.

24. Earlier explanatory models, only a few of which analyze *changes* in R&D/sales ratios, include Edwin Mansfield, *Industrial Research and Technological Innovation* (New York: Norton, 1968), chap. 2; the chapters by Richard C. Levin and Peter C. Reiss, Ariel Pakes and Mark Schankerman, John T. Scott, Jacques Mairesse and Alan K. Siu, and Mark Schankerman and M. Isaq Nadiri in Zvi Griliches, ed., *R&D, Patents, and Productivity* (Chicago: University of Chicago Press, 1984); and Charles P. Himmelberg and Bruce C. Petersen, "R&D and Internal Finance: A Panel Study of Small Firms in High-Tech Industries," working paper, Federal Reserve Bank of Chicago, 1990.

the year-to-year change in price-cost margins ΔPCM, anticipating a positive relationship.

During the 1970s and 1980s, special import quotas, tariffs, and other trade barriers were emplaced by the United States government with increased frequency. Their presence recalls the duality of theoretical results reviewed in Chapter 2. If competition spurs innovation, protection from import competition through trade barriers could lead companies to relax their R&D efforts. But if competition becomes so intense that it undermines the profitability of R&D, trade barriers might, by reducing import penetration, permit companies to support more intense technological effort. Our measure of the degree to which companies were protected began with a tabulation of affirmed trade restraint actions by four-digit industry. A dummy variable was coded to have unit value for each year during which Section 201 "escape clause" barriers or negotiated voluntary restraint agreements were in effect, and a second dummy had unit value for the first three years of restraint under other sections of the applicable U.S. Trade Act. These industry dummies were linked to the company level using the sales share weights w_{ijt}, and the two weighted average variables computed in this manner were summed to form the composite company trade protection index $BARS_{it}$.[25] In principle, $BARS$ could have a maximum value of 2, but no company had all of its lines protected in every year, and the mean value of $BARS$ was 0.135, with a standard deviation of 0.249. For the 53 percent of observations on companies with at least some protection in a given year, the mean value of $BARS$ (approximating the fraction of company sales protected) was 0.253.

Some research and development outlays reported to the Census Bureau as company-financed were actually oriented, Frank Lichtenberg has shown, toward winning future government R&D and procurement contracts.[26] We test for this phenomenon by including a variable $\Delta FEDRD$, measuring year-to-year changes in a company's federal contract R&D outlays as a percentage of total domestic sales. To capture the anticipatory character of the Lichtenberg hypothesis, a forward lag is used; that is, changes in company-financed R&D outlays from, say, 1978 to 1979 are related to changes in federal contract R&D outlays from 1979 to 1980. Because of data gaps, regressions including $\Delta FEDRD(T+1)$ had 29 fewer observations than those without.

---

25. See also Table 4.2.

26. See Frank Lichtenberg, "The Private R&D Investment Response to Federal Design and Technical Competitions," *American Economic Review*, 78 (June 1988), pp. 550–559.

Finally, one might expect companies headed by executives educated in science or engineering to defend their turf against high-technology competition from abroad more aggressively than those led by non-technical managers. This hypothesis is tested using the educational background variable *TECH*, described in connection with the Table 5.1 analysis of R&D/sales levels.

Table 5.2 provides an overview of the principal time series analysis variables, along with information on means and standard deviations.

Table 5.2   Principal time series variables[a]

Variable	Description	Mean	Standard deviation
$\Delta ADJRD$	Year-to-year change (from $T-1$ to T) in index-adjusted R&D/sales percentages	.033	.752
$\Delta ADJIMP(T)$	Change from $T-1$ to T in index-adjusted imports/output value percentages	.692	2.104
$\Delta ADJNX(T)$	Change from $T-1$ to T in index-adjusted net exports/output value percentages	−.501	3.302
$\Delta ADJIMP(TD)$	Triangularly distributed four-year lagged change in imports/output value percentages	.675	1.458
$\Delta ADJNX(TD)$	Triangularly distributed four-year lagged change in net exports/output percentages	−.496	2.172
ΔPCM	Change from $T-1$ to T in price-cost margin ratio	−.00032	.068
$\Delta FEDRD(T+1)$	Change from T to $T+1$ in federal contract R&D/sales percentage	−.024	.991
BARS	Approximate fraction of company sales covered by import barriers	.135	.249
$BUSCON(T)$	Business conditions index: percentage change in real GNP, $T-1$ to T	1.315	2.781

a. The number of observations is 4,790 except for $\Delta FEDRD(T+1)$ and *TECH*, for which $N = 4,761$ and 3,431, respectively.

Time Series Analysis Results

Table 5.3 presents the results of ordinary least squares regression analyses, with year-to-year changes in R&D/sales percentages as the dependent variable. Regressions using $\Delta ADJIMP$ as the import impact variable are presented in the left-hand columns and those using $\Delta ADJNX$ are in the right-hand columns.

Perhaps the most striking result is the weak explanatory power of the regressions, shown by low R^2 values. Tests for autocorrelation revealed that the considerable amount of residual "noise" was essentially "white," that is, without regular year-to-year cycles or trends.

Despite the high noise levels, systematic signals were detected. If we first consider regressions (3.1) and (3.6), the import reaction coefficients suggest weakly that on average, U.S. companies' R&D spending was cut back in response to positive import shocks, largely in the contemporaneous year. The coefficients for the net exports variable, which among other things strips out intra-company trade within multinational enterprises, reveal a stronger positive response distributed over the contemporaneous and preceding years. For the import variable, coefficients with negative signs connote "submissive" reactions, as articulated theoretically in Chapter 2; those with positive signs signify aggressive responses. For net exports, whose value falls with rising imports, a positive sign implies a *submissive* reaction. The results thus point preliminarily in the direction of submissive reactions. If rising imports had merely eroded companies' domestic sales without inducing R&D spending changes, the denominator of R&D/sales ratios would have fallen, driving the ratios upward, not downward, as the results indicate. A change in behavior consistent with the theory of submissive reactions is implied.

Regressions (3.2) and (3.7) add a forward import competition lag (requiring a substantial reduction in sample size). The coefficients fall far short of statistical significance, providing no indication that companies anticipated import shocks in their R&D spending decisions.

The use of forward-lagged federal contract R&D/sales ratio changes to test the Lichtenberg hypothesis was uniformly unsuccessful, perhaps because companies' contract-seeking entailed forward lags longer than the one year allowed by our data set, or because company-financed R&D outlays were incurred to win production contracts as well as R&D contracts. In most subsequent analyses, $\Delta FEDRD(T+1)$ is omitted to take advantage of all usable observations.

Regressions (3.3) and (3.8) compress the import and net export vari-

Table 5.3   Regression analysis of annual R&D/sales first differences:
308 companies, 1971–1987[a]

Explanatory variable	Regression number				
	(3.1)	(3.2)	(3.3)	(3.4)	(3.5)
$\Delta ADJIMP(T)$	−.0085 (1.60)	−.0091 (1.69)	—	—	—
$\Delta ADJIMP(T-1)$	.0059 (1.10)	.0099 (1.76)	—	—	—
$\Delta ADJIMP(T-2)$	−.0078 (1.40)	−.0024 (0.42)	—	—	—
$\Delta ADJIMP(T-3)$	−.0014 (0.25)	.0098 (1.60)	—	—	—
$\Delta ADJIMP(T+1)$	—	−.0015 (0.29)	—	—	—
$\Delta ADJIMP(TD)$	—	—	−.0093 (1.20)	.0017 (0.19)	−.0170 (1.50)
ΔPCM	−1.079 (6.69)	−.994 (6.31)	−.987 (6.23)	−.998 (6.30)	−1.311 (6.28)
$\Delta FEDRD(T+1)$	.0027 (0.25)	−.0003 (0.03)	—	—	.0014 (0.13)
$BARS$	−.0654 (1.49)	−.0704 (1.61)	−.0597 (1.37)	—	—
$\Delta ADJIMP(TD)$ × $BARS$	—	—	—	−.0583 (2.47)	—
$\Delta ADJIMP(TD)$ × $TECH$	—	—	—	—	.0156 (1.60)
16 year intercepts	s[b]	s	s	s	s
R^2	.0447	.0462	.0415	.0423	.0519
N	4,761	4,455	4,790	4,790	3,431

a. *T*-ratios are given in subscripted parentheses.
b. s: Coefficient suppressed.

Table 5.3   (continued)

Explanatory variable	Regression number				
	(3.6)	(3.7)	(3.8)	(3.9)	(3.10)
$\Delta ADJNX(T)$	.0099 (2.53)	.0088 (2.24)	—	—	—
$\Delta ADJNX(T-1)$	.0071 (1.83)	.0030 (0.73)	—	—	—
$\Delta ADJNX(T-2)$	.0004 (0.09)	−.0025 (0.59)	—	—	—
$\Delta ADJNX(T-3)$	−.0011 (0.27)	−.0059 (1.33)	—	—	—
$\Delta ADJNX(T+1)$	—	.0016 (0.41)	—	—	—
$\Delta ADJNX(TD)$	—	—	.0218 (4.02)	.0153 (2.57)	.0194 (2.14)
ΔPCM	−1.021 (6.34)	−.888 (5.58)	−.975 (6.16)	−.987 (6.24)	−1.140 (5.41)
$\Delta FEDRD(T+1)$	.0086 (0.79)	.0054 (0.46)	—	—	.0080 (0.72)
$BARS$	−.0545 (1.25)	−.0636 (1.47)	−.0586 (1.34)	—	—
$\Delta ADJNX(TD)$ × $BARS$	—	—	—	.0511 (2.70)	—
$\Delta ADJNX(TD)$ × $TECH$	—	—	—	—	−.0093 (1.30)
Intercepts	s	s	s	s	s
R^2	.0446	.0460	.0444	.0455	.0512
N	4,734	4,455	4,790	4,790	3,407

ables into a more parsimonious triangular lag structure with weights of 0.6 for year T, 0.25 for $T-1$, 0.10 for $T-2$, and 0.05 for $T-3$. This specification avoids the sign reversal problems that commonly intrude when several intercorrelated right-hand side variables are introduced together. Compared with their unconstrained four-lag counterparts, the triangular lag coefficients have either superior or insignificantly inferior explanatory power. Submissive reactions are again implied. A ten percentage point increase in imports as a percentage of domestic output value is found to reduce the average company's R&D/sales performance by just short of one-tenth percentage point, for example, from 3.22 percent (the all-company average) to 3.13 percent, all else equal. A ten percentage point *decrease* in net exports reduces the R&D/sales percentage average from 3.22 to 3.00 percent.

In all regressions, changes in company price-cost margins ΔPCM have signs contrary to the original expectation that higher profits would induce enhanced R&D. Further investigation clarified this seeming mystery. More than half of all industrial R&D employees work outside free-standing laboratories or other central offices; that is, they are employed within plants that also produce goods for sale.[27] An increase in R&D outlays thus raises in-plant materials and payroll costs, reducing price-cost margins in the short run, consistent with the negative ΔPCM coefficients. Rewards in the form of increased PCMs presumably follow only with a lag.

The negative and marginally significant coefficients on *BARS* suggest that companies whose industries enjoyed special import protection had slightly lower R&D/sales ratio increases, all else (including changes in import penetration) held equal.[28] Regressions (3.4) and (3.9) take a further step, exploring whether protection from imports affected the strength of companies' *reactions* to rising import competition. For both imports and net exports, the interaction effects are statistically significant, but because of multicollinearity, they erode the non-interacted reaction coefficient (and for imports, force a sign reversal). For a hypothetical company protected continuously from import competition in all of its lines (*BARS* = 1), a ten percentage point increase in imports would lead to a predicted R&D/sales drop of 0.57 percentage points. A similar decline in net exports with full protection implies an R&D/sales drop of 0.66 percentage points. Strong protection thus appears to have made companies' reactions more submissive on average.

---

27. U.S. Bureau of the Census, *Enterprise Statistics: 1982,* vol. 2, "Auxiliary Establishment Reports" (Washington, D.C.: U.S. GPO, December 1986), pp. 4, 11.

28. When the *BARS* variable was divided into its two constituent components, their coefficients had the same signs and similar magnitudes. The coefficients for "escape clause" actions, which tend to cover more of an industry's output, were larger. But because few industries had such protection, they had lower *t*-ratios.

To be sure, the presence of trade barriers might have reversed a rising tide of imports. And if imports *fell,* the implication of regressions (3.4) and (3.9) is that R&D/sales ratios should have risen appreciably— plausibly, a desired consequence of protection. However, this is belied by the negative coefficients on *BARS* in the non-interactive regressions. More important, trade barriers did not succeed in reducing the growth of imports below the average (positive) rates experienced by all companies. The zero-order correlation between $\Delta ADJIMP(T)$ and $BARS(T)$ was 0.0001. Thus, with imports rising on average as much with protection as without, the characteristic reaction of protected companies was more submissive than for companies lacking protection. The observed tendency for imports to rise as much with protection as without it may mean that protection stabilized or reduced import quantities, but triggered offsetting increases in the prices at which imported goods were sold.

Regressions (3.5) and (3.10) introduce an interaction between the triangularly lagged import variables and the dummy variable *TECH,* indicating whether a company had a leader educated in science or engineering.[29] For both imports and net exports, there is a significant increase in R^2 values relative to unreported regressions lacking the educational background variables. Having a technically educated top executive appears to have made companies more aggressive on average. Indeed, in regression (3.5), the predicted consequence of a ten percent increase in imports for companies with a technically educated leader is an R&D/sales decrease of only 0.014 percentage points. Or in regression (3.10), the predicted impact of a ten percent decrease in net exports is 0.10 percentage points—less than half the magnitude estimated in regression (3.8).

During the late 1970s and mid-1980s, we have seen, R&D/sales ratios in our sample of companies (and in American industry more generally) exhibited distinct upward trends. The regressions used thus far control for those trend effects (as well as business cycle effects) by means of annual dummy variables. Table 5.4 explores what happens when only business cycle effects are controlled using three lagged terms of the variable *BUSCON.* The *BUSCON* indices have the anticipated negative signs, showing the tendency of R&D outlays to fall more slowly than sales in a recession and to rise less slowly than sales in an upturn. The

---

29. The number of observations drops sharply because, as indicated previously, the educational background variables could be quantified for only 221 of our 308 sample companies.

Table 5.4   Time series regressions without year controls ($N = 4,790$)[a]

Explanatory variable	Regression number					
	(4.1)	(4.2)	(4.3)	(4.4)	(4.5)	(4.6)
$\Delta ADJIMP(TD)$	−.0006	.0085	−.0343	—	—	—
	(0.08)	(0.97)	(0.28)			
$\Delta ADJNX(TD)$	—	—	—	.0098	.0039	−.244
				(1.91)	(0.68)	(2.39)
ΔPCM	−.986	−.995	−.986	−.980	−.990	−.970
	(6.20)	(6.26)	(6.20)	(6.16)	(6.22)	(6.10)
$BARS$	−.032	—	−.033	−.030	—	−.026
	(0.74)		(0.75)	(0.69)		(0.60)
$BUSCON(T)$	−.032	−.033	−.033	−.031	−.031	−.031
	(7.07)	(7.13)	(7.07)	(6.63)	(6.65)	(6.64)
$BUSCON(T-1)$	−.005	−.005	−.005	−.004	−.004	−.004
	(1.22)	(1.22)	(1.22)	(1.03)	(1.05)	(0.93)
$BUSCON(T-2)$	−.011	−.011	−.011	−.011	−.011	−.009
	(2.52)	(2.62)	(2.53)	(2.60)	(2.62)	(2.09)
$\Delta ADJIMP(TD)$ × $BARS$	—	−.048	—	—	—	—
		(2.04)				
$\Delta ADJNX(TD)$ × $BARS$	—	—	—	—	.045	—
					(2.37)	
$\Delta ADJIMP(TD)$ × $YEAR$	—	—	.0004	—	—	—
			(0.28)			
$\Delta ADJNX(TD)$ × $YEAR$	—	—	—	—	—	.0031
						(2.49)
Intercept	.097	.092	.097	.099	.096	.101
	(6.30)	(6.33)	(6.31)	(6.55)	(6.77)	(6.71)
R^2	.0223	.0231	.0223	.0231	.0241	.0243

a. *T*-ratios are given in subscripted parentheses.

contemporary (year T) effect is the strongest. With no other control for time trends, the import reaction coefficients fall—to nearly zero in imports regression (4.1) and by more than half in net exports regression (4.4) relative to otherwise comparable regression (3.8). The trade barrier effects and their interactions (the latter in regressions (4.2) and (4.5)) are also weakened. If rising import penetration caused many American companies gradually to realize that they had to intensify their R&D to remain competitive, controlling for trend effects through annual dummy variables, as is done in Table 5.3, biases measured reaction coefficients in a submissive direction. Omitting trend controls, as in Table 5.4, avoids any such bias. However, if R&D/sales ratios rose generally for other reasons—for example, because stimulative tax credits were put in place beginning in 1981, or because for some reason companies faced richer technological opportunities—the *omission* of trend controls biases the results in the opposite direction. It is impossible to be confident which chain of causation, and hence which specification, is correct. Regressions (4.3) and (4.6) in Table 5.4, which include import and net export terms with a calendar year interaction effect, do not resolve the puzzle. Reactions are found to move slightly but statistically insignificantly in a more aggressive direction over time in the imports regression, but toward significantly greater submissiveness in the net exports regression. We return to this problem from a different perspective subsequently.

To sum up what we have learned thus far, the time series analysis reveals much unsystematic variation in company R&D/sales ratio changes from year to year and a weak average tendency toward submissive reactions to rising import competition, at least in the short run. Having protection from imports appears to have made company reactions more submissive on average. The presence of a top executive educated in science or engineering nudged companies in the direction of more aggressive responses.

Why Reactions Differ

The case studies in Chapter 3 illustrate widely varying company reactions—sometimes aggressive, sometimes submissive—to new high-technology competition from abroad. In view of this, it is reasonable to ask whether the regressions in Tables 5.3 and 5.4 have little explanatory power in part because firms' quantitative reactions to import shocks were heterogeneous. In this section we provide support for that hypothesis and explore why reactions differed from one company to another.

For each of the 308 sample members, we computed individual time series regressions of the form:

$$[5.6] \quad \Delta ADJRD_{it} = a + b_1 \Delta ADJIMP(TD)_{it} + b_2 YEAR + b_3 BUSCON_{it} + e_{it},$$

where $\Delta ADJIMP(TD)$ is a triangularly distributed import lag similar to those tested in regressions (3.3)–(3.5) (replaced by $\Delta ADJNX(TD)$ in other individual company regressions), *YEAR* is a time trend index, and *BUSCON* is the business cycle index emphasized in Table 5.4. *BUSCON* and *YEAR* together control for business cycle and trend effects more parsimoniously than annual dummy variables. By including *YEAR,* we emphasize short-run reactions, in effect assuming that the attenuation of import reactions observed in Table 5.4 was driven by exogenous trend effects. Whatever the exact causality, the trend effect is not strong. In an analogue of Table 5.4's regression (4.1) with *YEAR* added, the positive *YEAR* coefficient's *t*-ratio was 4.19. *YEAR* coefficients were positive in 60 percent of the 308 individual company $b_{\Delta ADJIMP}$ regressions and 56 percent of the $b_{\Delta ADJNX}$ regressions. Inter-company differences in the *YEAR* and *BUSCON* variables were not statistically significant, with *F*-ratios of 1.03 and 1.07 for the depooling of import and net export regressions, respectively, containing *YEAR, BUSCON,* and individual company intercepts.

Adding 308 company-specific import reaction coefficients to the regressions of $\Delta ADJRD$ on *BUSCON* and *YEAR* did reveal significant heterogeneity, with $F(308,3558) = 1.37$ for the triangularly distributed import regressions and $F(308,3531) = 1.46$ for the net export regressions.[30] Taking into account these individual company effects raises the proportion of variance in R&D/sales ratio changes explained (that is, implied R^2 values) to 0.293 and 0.305 for imports and net exports, respectively. The mean import reaction coefficient value was -0.088—nine times the value estimated in Table 5.3's pooled regression (3.3). The average value of the five coefficients nearest the median was -0.032.[31] Fifty-nine percent of the 308 import reaction coefficients were negative. The mean net export reaction coefficient value was $+0.021$, nearly the same as the value estimated in regression (3.8). Positive coefficients emerged for 55.5 percent of the 308 companies. The disaggregated regressions thus continue to exhibit submissive reactions on average, but with marked heterogeneity.

Companies might react differently to intensified import competition because of dissimilar sales and market structures, technological opportunities advancing at unequal rates, diverse means of capturing the benefits

---

30. The 99 percent confidence point in an *F*-test is 1.24.
31. Census confidentiality rules preclude the disclosure of values for individual companies, including single-observation median values.

from technological innovation, and more or less rich links to innovation-facilitating science bases. The role of these differences is investigated in two stages.

Some of the 308 sample companies operated mainly in the United States; others had extensive multinational activities. We test the role of multinationality by identifying the subset of companies reporting R&D expenditures outside the United States in any Census Bureau survey year.[32] The average reaction coefficients b_1 (from equation [5.6] above) for the two groups, with standard errors in parentheses, were as follows:

	Imports	Net exports
191 companies with overseas R&D	−0.097	−0.003
	(0.039)	(0.019)
117 companies without overseas R&D	−0.074	+0.061
	(0.032)	(0.026)

For companies without overseas R&D, the mean reactions were consistently submissive and of similar magnitude. Companies *with* overseas R&D differed insignificantly from those without in their reaction to changing imports; $F(1,306) = 0.16$. However, the two cohorts reveal quite different mean reactions to changes in net exports; $F(1,306) = 3.98$, which exceeds the 95 percent confidence point of 3.88. On average, R&D multinationals exhibited a near-zero reaction to changes in net exports—a variable that nets out intra-company imports and exports, and which may therefore measure foreign competition more meaningfully than the import variable. The reactions of R&D multinationals to changes in imports, without adjustment for offsetting exports, were also more heterogeneous than those of domestic specialists; that is, they exhibited wider statistical dispersion. The null hypothesis of homogeneous import reaction coefficient variances is rejected at the 95 percent confidence level, with $F(191,117) = 2.39$.[33] The widely varying import reactions of R&D multinationals may mean that the import variable measures threats of diverse potency (for instance, with rising imports coming in some cases from rivals and in others from companies' own overseas branches). Or R&D

---

32. The "any year" criterion was applied because non-response rates were higher on such questions than on domestic R&D spending questions.

33. For the variances of net export reaction coefficients, $F = 1.13$, which is not statistically significant.

multinationals may be able to increase defensive R&D overseas as well as, or instead of, in the United States.

The impact of other environmental variables on company reactions is tested by estimating cross-sectional regression equations of the form:

[5.7] $b_{i,\Delta ADJIMP(TD)} = m + k\mathbf{Z}_i + e_i,$

where $b(\cdot)$ is the import or net exports reaction coefficient estimated from individual company regressions [5.6] above and $\mathbf{Z}$ is a matrix of explanatory variables.[34] Among the explanatory variables, several possibilities are potentially relevant.

A given percentage increase in an industry's import penetration reduces all producers' domestic sales more or less proportionately on average, but it causes larger absolute sales (and presumably, quasi-rent) losses for firms with large market shares than it does for smaller sellers. As the analysis underlying Figure 2.7 implies, firms with larger market shares and hence larger absolute losses are expected to react more aggressively, all else equal. We measure this structural influence through three variables: *CR4,* the average four-seller 1977 domestic concentration ratio for the industries occupied by a company, weighted as in other industry–company links by the firm's sales share w_{ijt},[35] *LOGSALES,* the logarithm (to base 10) of average company sales in the United States over the 17-year sample period; and *DIVERS,* an index of diversification obtained by summing the squared values of w_{ijt} and averaging those values over the sample period.[36] The more specialized a company was in one industry, the more nearly *DIVERS* approaches unity from its lower bound of zero.[37]

Companies' multinationality is represented by the variable *OVERSEAS,* which is the average across all years with data of R&D spending in subsidiaries outside the United States as a percentage of U.S. sales. The mean of *OVERSEAS* was 0.284 percent, with 117 companies having zero values.

---

34. Because the $b(\cdot)$ coefficients are estimated with varying precision, regression [5.7] is computed using weighted least squares, with the inverse of the $b(\cdot)$ coefficient standard errors serving as weights.

35. It is scaled in percentage terms, with a mean value of 40.4.

36. For an earlier use of *DIVERS* (or more precisely, its inverse) as an index of diversification, see F. M. Scherer and David Ross, *Industrial Market Structure and Economic Performance* (third ed.; Boston: Houghton-Mifflin, 1990), pp. 91–92.

37. The mean value of *DIVERS* was 0.474, with a standard deviation of 0.270.

To determine how important alternative methods of capturing the quasi-rents from innovation were perceived to be and the strength of company links to the science base, we tap again, as in Chapter 4, the Yale University survey of 650 industrial R&D managers.[38] The survey responses are available for 130 Federal Trade Commission Line of Business four-digit reporting categories, including most of the industries in which our sample companies concentrated their efforts. Potentially relevant variables from the Yale survey data set were linked to the sample companies using the weighting factor w_{ijt}, with $t = 1979$.[39] Table 5.5 lists mnemonics, descriptions, means, and standard deviations of the Yale variables used in the present analysis. The variables' standard deviations here are lower than in the industry cross-section analysis of Chapter 4, because some averaging-out of disparate values occurred with aggregation to the company level. Some of the variables are also intercorrelated, causing estimated coefficient values to be unstable when several collinear variables were introduced together. We therefore proceed cautiously, adding the variables in clusters. Table 5.6 reports the principal results, with import reaction coefficient regressions on the left-hand side and net export coefficient regressions on the right-hand side.

Consistent with the dichotomous results reported previously, net export reactions were less submissive in companies with R&D outside the United States, although the coefficients for continuous overseas R&D/ sales ratios fall short of statistical significance.[40]

The structural hypotheses are strongly supported, although the exact chain of causation is left in doubt because of collinearity among the *LOG-SALES, DIVERS,* and *CR4* variables. (The simple correlation between *DIVERS,* whose value falls with greater diversification, and *LOGSALES*

---

38. See Richard C. Levin, Alvin K. Klevorick, Richard R. Nelson, and Sidney G. Winter, "Appropriating the Returns from Industrial Research and Development," *Brookings Papers on Economic Activity,* 1987, no. 3, pp. 783–820.

39. The weighted Yale survey variables are highly correlated from year to year. Thus, $r_{72,79}$ for 308 companies and all Yale variables used in our analysis is 0.988, and $r_{79,85} = 0.978$.

40. When separate regressions were estimated for companies with positive and zero *OVERSEAS* values, Chow tests for global regression coefficient heterogeneity were rejected, with F-ratios ranging from 0.41 to 0.61. Although R&D multinationals reacted differently to import shocks, their reactions were influenced similarly by the structure, appropriability, and science base variables. A dummy variable identifying seven companies owned by foreign parents throughout the 1971–1987 sample period had signs identical to those for U.S.-based multinationals, but its coefficients were not statistically significant.

Table 5.5   Variables from the Yale survey[a]

Variable	Survey question paraphrase	Mean	Standard deviation
PRODPAT	How effective are patents as a means of capturing and protecting the advantages from new or improved products?	4.36	.74
LEARNING	How important is moving quickly down the learning curve as a means of capturing and protecting the advantages from new or improved products?	5.20	.41
SERVICE	How important are superior sales or service efforts as a means of capturing and protecting the advantages from new or improved products?	5.62	.42
SCIENCE	How relevant were the basic sciences of biology, chemistry, and physics (average of three) to technological progress in this line of business over the past 10–15 years?	3.93	.55
NICHES	To what extent have technological activities been oriented toward designing products for specific market segments?	5.17	.45
PCTPROC	What percentage of total R&D in the industry is directed toward new production processes, as distinguished from new and improved products? (Scaled from 0 to 100)	27.4	21.3

a. Measured on a Likert scale of 1 to 7, with "7" implying "very effective" or "very important" or "very relevant" and 1 the opposite, unless otherwise stated.

was -0.517; the correlation between *CR4* and *LOGSALES* was $+0.096$.) *LOGSALES* has higher *t*-ratios than *DIVERS*, but undermines the statistical significance of *CR4*, which, without *LOGSALES*, is consistently significant. *DIVERS* shares a significant explanatory role with *CR4*, with which it was uncorrelated ($r = +0.036$), in both regressions. Evidently, some combination of large firm size, high seller market shares, and extensive diversification made firms' R&D reactions to import shocks more aggressive. Whether diversification had its impact because it implies larger size, even though smaller market shares for a given sales volume,

Table 5.6 Regressions of company import reaction coefficients on explanatory variables (308 companies)[a]

Explanatory variable	Dependent variable: $b_{\Delta ADJIMP(TD)}$				Dependent variable: $b_{\Delta ADJINX(TD)}$			
	(6.1)	(6.2)	(6.3)	(6.4)	(6.5)	(6.6)	(6.7)	(6.8)
OVERSEAS	−.004 (0.14)	−.007 (0.21)	−.014 (0.43)	.013 (0.40)	−.019 (0.88)	−.016 (0.73)	−.015 (0.70)	−.021 (0.96)
LOGSALES	.058 (2.87)	—	—	—	−.045 (3.09)	—	—	—
DIVERS	—	−.099 (2.04)	−.099 (2.11)	−.093 (2.01)	—	.062 (1.87)	.062 (1.93)	.059 (1.86)
CR4	.0009 (1.49)	.0017 (2.55)	.0013 (2.11)	.0021 (3.18)	−.0005 (1.00)	−.0011 (2.28)	−.0011 (2.28)	−.0011 (2.37)
PRODPAT	−.047 (2.68)	−.041 (2.33)	—	−.053 (2.94)	.0056 (0.46)	.0057 (0.46)	—	.0065 (0.53)
SERVICE	−.023 (0.68)	−.038 (1.09)	−.032 (0.94)	−.043 (1.28)	.0031 (0.14)	.009 (0.40)	.010 (0.43)	.012 (0.54)
LEARNING	—	—	.032 (0.93)	—	—	—	.011 (0.47)	—
SCIENCE	−.057 (2.07)	−.049 (1.77)	−.051 (1.83)	−.067 (2.37)	.022 (1.16)	.018 (0.96)	.019 (0.99)	.022 (1.14)
NICHES	−.022 (0.88)	−.0019 (0.07)	—	—	.012 (0.71)	−.0022 (0.11)	—	—
PCTPROC	—	—	—	.0010 (2.03)	—	—	—	−.0003 (0.90)
Intercept	.27 (0.92)	.46 (1.43)	.18 (0.72)	.63 (2.85)	.10 (0.50)	−.18 (0.85)	−.16 (0.88)	−.14 (0.96)
R^2	.124	.135	.105	.135	.110	.084	.081	.080

a. T-ratios are given in subscripted parentheses. Generalized (weighted) least squares is used. The R^2 values are for unweighted regressions with identical variables.

or whether more diversified companies had greater shock absorption capacity and the resources to cross-subsidize lines under import attack, remains unclear.

The more aggressive reactions of producers in concentrated industries provide new support for the neo-Schumpeterian argument that monopoly (or more plausibly, oligopoly) power fosters vigorous innovation in markets subjected to the "creative destruction" of technological change.[41] However, the exact chain of causation merits further exploration. The *CR4* coefficients of Table 5.6 might be picking up theoretically plausible "fast second" reactions by dominant firms that lagged in being first movers on new products or processes.[42] It is also conceivable (and consistent with qualitative evidence[43]) that oligopolistic sellers pursued pricing and marketing policies that encouraged import inroads—for example, by holding their prices above entry-deterring levels.[44]

Statistical support for the latter hypothesis comes from regressions seeking to explain the import and net export experience of our 308 sample companies using seller concentration and profitability variables. A company's average import growth experience (*IMPGROWTH*) was measured as the coefficient obtained by regressing the company's annual weighted average imports as a percentage of manufacturing shipments value on a linear 1971–1987 time trend. Net export growth (*NXGROWTH*) was measured analogously. The average values of these annual growth coefficients were 0.64 and −0.42 percentage points respectively. These were regressed on 17-year average values of company price-cost margins (*AVPCM*) and the weighted average concentration indices (*CR4*). With 308 observations in each case, the resulting regressions were:

[5.8] $IMPGROWTH = -.04 + .0162CR4 + .098AVPCM,$
$$(0.18)   (4.23)   (0.20)$$

$$R^2 = 0.056$$

---

41. See Joseph A. Schumpeter, *Capitalism, Socialism, and Democracy* (New York: Harper, 1942), chaps. 7 and 8; and Scherer and Ross, *Industrial Market Structure*, chap. 17.

42. See Figure 2.7 and the related discussion.

43. See the testimony on automobiles, steel, and electrical motors by F. M. Scherer in U.S. House of Representatives, Committee on the Judiciary, Subcommittee on Monopolies and Commercial Law, Hearings, *Corporate Initiative* (Washington, D.C.: U.S. GPO, 1982), pp. 26–54.

44. See Scherer and Ross, *Industrial Market Structure*, especially pp. 370–371 and 396.

and

[5.9] $NXGROWTH = -.14 - .0087CR4 + .279AVPCM,$
             (0.64) (2.30) (0.58)

$R^2 = 0.020.$

Imports rose significantly, and net exports fell, with higher levels of concentration.[45] The *AVPCM* results are weak and (for *NX*) counter-intuitive. This might stem from the averaging of price-cost margins over all sample years, including early years, whose high price-cost margins stimulated increased import penetration, and later years, during which margins were squeezed as a consequence of import competition. In an attempt to disentangle the profit relations, year-to-year changes in adjusted imports were regressed on current and one-year-lagged *PCM* values in a pooled time series–cross section analysis:

[5.10] $\Delta ADJIMP(T) = .88 - .542PCM(T) - .129PCM(T-1),$
              (12.48) (1.18) (0.28)

$R^2 = 0.0018,$ $N = 4,790.$

The results weakly refute the hypothesis that high current price-cost margins induced imports. To the contrary, PCMs may have been eroded by rising imports. The effects of market structure are apparently captured directly by the *CR4* measure rather than by the pricing and profits that flowed from it.[46]

If higher seller concentration stimulated imports but simultaneously made firms alter their R&D more aggressively in response to increasing imports, what was the net effect on R&D? Consider the effects of a 40 point increase in concentration, for instance, from the sample mean of 40.4 to a value of 80.4, implying tight oligopoly. By regressions [5.8]

---

45. The results persisted, and *CR4* coefficients were of similar magnitudes, when additional variables measuring import barriers, the extent of overseas R&D, and diversification were added. The *DIVERS* coefficient was statistically significant; more specialized companies experienced greater import penetration.

46. Alternatively, imports may have risen because of high prices, most of the rents from which were captured not as profit but as premium wages. Compare Thomas A. Pugel, "Profitability, Concentration, and the Interindustry Variation in Wages," *Review of Economics and Statistics,* 62 (May 1980), pp. 248–253; Kim B. Clark, "Unionization and Firm Performance: The Impact on Profits, Growth, and Productivity," *American Economic Review,* 74 (December 1984), pp. 893–919; and Thomas Karier, "Unions and Monopoly Profits," *Review of Economics and Statistics,* 67 (February 1985), pp. 34–42.

and [5.9], average import growth would have been increased by 0.65 percentage points per year while net exports fell by 0.35 points. The median R&D reaction to a one percentage point increase in imports, measured from individual company regressions, was −0.032 percentage points. From regression (6.2) in Table 5.6, a 40 point *CR4* increase moves reactions to a one percentage point import increase in the aggressive direction by 40 × .0017 = 0.068, all else equal. The R&D-stimulating effect of concentration outweighs the R&D-retarding effect of increased imports by 0.068 − 0.032 = 0.036. With a 40 point difference in concentration raising imports by 0.65 points, the combined effect is 0.65 × 0.036, or an increase of 0.0234 in R&D as a percentage of sales. For net exports, the impact of a 40 point *CR4* increase is a change in the mean reaction coefficient value of 0.021 by −0.0011 × 40 = −0.044 points. Thus, reactions again move from submissive to aggressive, all else equal. This plus the tendency of concentration to *reduce* net exports leads to *increased* R&D/sales percentages—thus, by −0.35 × −.044 = +0.015 percentage points per year, all else equal. These estimates are sensitive inter alia to the assumed reaction coefficient values, which have sizable standard errors. If, instead of using the median import reaction coefficient estimate of −0.032, we substitute the mean value of −0.088, the net effect of a 40 point concentration ratio increase would be a reduction in R&D/sales percentages, not the increase found previously. A margin of uncertainty remains, although the net impact of sizable concentration increases was most likely R&D-enhancing.

For the Yale appropriability and science base variables, the results are mixed. Submissive reactions were associated with strong patent protection, although for net exports, the coefficients are not statistically significant. An emphasis on customer service was consistently but even more weakly linked to submission. When market positions were captured by moving rapidly down learning curves, aggressive reactions were somewhat more likely. (In unweighted net export regressions, *LEARNING* was statistically significant.) *SCIENCE* has coefficients significant for imports, but not net exports, whose signs contradicted our initial hypothesis that firms with strong science base links would react more aggressively. A possible rationalization is that in industries for which the basic sciences are particularly relevant, innovation-facilitating knowledge diffuses rapidly to foreign competitors as well as to domestic firms. And Japanese companies, which have mounted the most pervasive challenge to U.S. firms, have proved particularly adept at exploiting the spillover benefits generated by academic science.[47] An alternative technological opportunity measure (*PROGRESS* in Chapter 4), characterizing the rate at which new and improved products were introduced during the 1970s, was consistently insignificant. Companies occupying industries in which niche-filling was an important strategy exhibited weakly submissive reactions. For imports but not for net exports, regressions (6.4) and (6.8) reveal, reactions were significantly more aggressive in industries emphasizing process R&D, on which Japanese firms spend a substantially higher fraction of their R&D budgets than does American industry.[48] A variable measuring the extent to which Yale survey respondents stressed product standardization in their R&D (*STANDARD* in Chapter 4) had little explanatory power in either import or net export reaction coefficient regressions.

Finally, we confront the data with a question posed by limitations in our case study sample coverage. None of the case studies came from the large and technologically dynamic chemicals sector (SIC 28). Do the quantitative analyses provide any clues that companies at home in the chemicals industry reacted to import competition in a way different from the general patterns observed in Table 5.6? Dummy variables with unit values for 50 chemical industry companies were introduced into weighted regressions also containing the *CR4, DIVERS,* and *OVERSEAS* variables. Chemical industry members were found to react 0.13 points (t = 3.13) more submissively to increased imports, all else equal. However, their more submissive reaction to net export changes was small (0.017) and statistically insignificant (t = 0.52). As on several other questions, the answer is apparently sensitive to whether competition from abroad is measured in terms of imports alone or net exports, the latter netting out non-competitive intra-industry trade among the national units of chemical companies, many of which have extensive multinational operations.

---

47. See, for example, Paul E. Gray (former president of MIT), "Advantageous Liaisons," *Issues in Science and Technology,* Spring 1990, p. 43; Industrial Research Institute, Government-University Industry Research Roundtable, "Industrial Perspectives on Innovation and Interactions with Universities" (Washington, D.C.: National Academy Press, February 1991); "Picking Japan's Research Brains," *Fortune,* March 25, 1991, pp. 84–96; and "The Lab's Role in U.S. Growth," *New York Times,* May 24, 1991, p. D2.

48. See Edwin Mansfield, "Industrial R&D in Japan and the United States: A Comparative Study," *American Economic Review,* 78 (May 1988), p. 226; and David B. Audretsch and Hideki Yamawaki, "R&D Rivalry, Industrial Policy, and U.S.-Japanese Trade," *Review of Economics and Statistics,* 70 (August 1988), pp. 438–447.

Long-Run Growth Relationships

The reaction coefficient relationships estimated thus far were for relatively short periods, with lags of at most three years. It is conceivable that longer-run reactions were more aggressive on average, for instance, as companies defeated in one round of a new product competition strained to catch up when the next generation of products was developed. We have seen hints of this in Table 5.4's finding that reaction coefficients were less submissive when time trends were left uncontrolled. For further insight, we analyze a variable *RDGROWTH*, which measures the average percentage rate per year at which a company's R&D/sales ratios grew over the 1971–1987 time frame. It was estimated by regressing the logarithms of (RD/S) on a calendar year variable. Its average value for the 308 companies was 1.87 percent, with a standard deviation of 4.72 percent.

We expect the growth of R&D intensity to be influenced not only by the vigor of import competition but also by company structure and technological opportunity variables. Some are drawn from the Yale survey, others from Census data. They are listed, along with brief definitions, means, and standard deviations, in Table 5.7. Most have transparent theoretical rationales. *RDEVIATE* is less obvious. It accounts for the possibility of Galtonian regression. That is, firms which are R&D over-achievers relative to averages of the industries they occupy might be expected to have relatively low R&D growth rates, all else equal, while under-achievers might have higher growth rates.

Table 5.8 reports the results of weighted least squares regressions for most of the variables defined in Table 5.7. Each observation is weighted by the inverse of the standard error from the regression estimating that company's R&D/sales growth rate. The import and net export growth variables, which are highly correlated ($r = -0.90$), are introduced separately.

The import competition variables imply a long-run tendency toward aggressive reactions.[49] In this respect they go beyond the insights yielded by Tables 5.3 and 5.4, which displayed only reduced submissiveness when time trends were left uncontrolled, especially for imports, but not a transition to aggressive reactions. However, the new long-run trend

---

49. Weakly aggressive innovation responses to import growth between 1976 and 1983 are reported for 26 West German industries by Joachim Wagner and Lutz Bellmann in "Produkt- und Prozessinnovationen als Unternehmensstrategien bei Importdruck," *Ifo-Studien*, 33 (1987, no. 3), pp. 223–242.

Table 5.7 Variables used in the analysis of R&D growth rates

Variable	Description	Mean	Standard deviation
IMPGROWTH	Annual trend change in imports as a percentage of domestic output value, 1971–1986 (percent)	.64	1.02
NXGROWTH	Annual trend change in net exports as a percentage of domestic output value, 1971–1986 (percent)	– .42	.99
PROGRESS	Rate of product technology advance during the 1970s (from Yale survey)	4.60	.71
FUTPROG	Yale survey variable recording whether product development opportunities were expected to be greater in 1980s than in 1970s	5.02	.42
AVBARS	Seventeen-year average of Table 5.2's trade barriers variable	.130	.151
AVPCM	Average price-cost margin in the company's domestic manufacturing operations, 1971–1986	.275	.118
AVFEDRD	Federally supported R&D as a percentage of domestic sales; 1971–1987 average	.76	2.70
OVERSEAS	R&D conducted in company laboratories outside the United States as a percentage of U.S. sales	.28	.63
FOREIGN	Dummy variable with unit value for companies owned by a foreign parent throughout 1971–1987	s[a]	s
RDEVIATE	Arithmetic difference between actual company-financed R&D/sales ratios, averaged over 17 years, and the similarly averaged *RDINDEX* values	.84	1.96

a. s: Suppressed.

Table 5.8   R&D growth rate regressions (308 companies)[a]

Independent variables	Regression number	
	(8.1)	(8.2)
IMPGROWTH	.306	—
	(1.25)	
NXGROWTH	—	− .165
		(0.64)
AVBARS	− 2.84	− 2.84
	(1.81)	(1.81)
AVPCM	4.07	4.00
	(2.25)	(2.21)
FUTPROG	1.12	1.15
	(2.07)	(2.12)
AVFEDRD	.075	.076
	(0.99)	(0.99)
OVERSEAS	.415	.403
	(1.24)	(1.20)
FOREIGN	− .466	− .464
	(0.34)	(0.34)
RDEVIATE	− .104	− .093
	(0.90)	(0.81)
Intercept	− 5.04	− 5.04
	(1.80)	(1.80)
R^2	.075	.070

a. *T*-ratios are given in subscripted parentheses. The technique is weighted least squares, but the R^2 values are for unweighted regressions with identical variables.

coefficients fall short of conventionally accepted statistical significance thresholds, so we are left in doubt as to the potency of the long-run change.[50] The erection of trade barriers continues to be accompanied by slower R&D growth, all else held equal. Whether this reflects the lulling

---

50. Substituting for the *OVERSEAS* variable *IMPGROWTH* × *DUMMY* and *NXGROWTH* × *DUMMY* interaction terms, where *DUMMY* took on unit values for companies with overseas R&D and zero otherwise, yielded interaction coefficients far short of statistical significance. The overall goodness of fit also deteriorated. Introducing quadratic *IMPGROWTH* and *NXGROWTH* terms showed no evidence of nonlinearities.

effect of protection or the possibility that, despite protection, import growth undermined the profitability of R&D cannot be inferred confidently. That both import growth and profitability are taken into account by other variables lends support for the "lulling" interpretation.

Higher price-cost margins were associated with more rapid R&D growth rates as well as with higher levels of R&D (compare Table 5.1). Given that much of the growth in R&D/sales ratios for U.S. manufacturers occurred during the 1980s, the strong performance of *FUTPROG*, predicting whether technical change was expected to accelerate during that period, is not surprising. The *PROGRESS* variable, evaluating the pace of product technology change during the 1970s, had no explanatory power and is omitted from the Table 5.8 regressions.[51] Consistent with the findings of Lichtenberg,[52] an active position in federal R&D contracting was weakly conducive to the growth of company-financed R&D during our sample period, with rapidly rising military R&D and procurement outlays in the late 1970s and early 1980s. The establishment of R&D laboratories overseas does not appear to have impaired the growth of R&D spending at home. U.S. R&D operations owned by foreign corporations experienced lower growth rates, although the effect falls short of statistical significance. There is only weak evidence of Galtonian regression in R&D growth rates for companies with above- and below-average R&D/sales ratio levels.

Conclusion

Using unusually rich data collected by the U.S. Census Bureau, we have analyzed the R&D spending reactions of U.S. companies to high-technology import competition, which hit U.S. companies with increasing force between 1971 and 1987. Most year-to-year changes in company R&D/sales ratios were unsystematic, related neither to import competition movements nor to other plausible explanatory variables. As in the case studies, we find considerable idiosyncrasy in companies' competi-

---

51. Also lacking explanatory power was the *SCIENCE* variable used in Table 5.6 and in the intra-industry trade analysis of Chapter 4. In another test, chemicals industry members were found to have R&D growth rates 0.42 to 0.47 points more rapid than other sample members, ceteris paribus, but the coefficients' *t*-ratios were only in the range of 0.59 to 0.67. Other coefficients remained essentially unchanged when chemicals industry dummy variables were added.

52. See note 26.

tive reactions. But import competition does appear to have made a difference. The short-run reaction to increased imports (or declining net exports) was on average submissive; that is, R&D/sales ratios fell. However, large, diversified firms occupying concentrated markets reacted more aggressively than their smaller, less diversified counterparts. Multinationals reacted more heterogeneously to rising imports and less submissively to net export declines. Insulation from import competition through trade barriers blunted firms' short-run reactions. Over the longer run, there appears to have been a reversal of the average reaction pattern from submissive to aggressive, although the evidence on this point remains weak.

6   Conclusions

Now, *here*, you see, it takes all the running you
can do, to keep in the same place. If you want to
get somewhere else, you must run at least twice as
fast as that!

—Lewis Carroll

In the looking-glass world of international trade, sustaining comparative
advantage in the supply of technologically advanced products takes a lot
of running. For such products, comparative advantage comes not from
Providential endowments but from the aggressive pursuit of technical
superiority and risk-taking. In other words, it requires Thomas Edison's
peculiar brand of genius—one percent inspiration and ninety-nine percent
perspiration. As more and more nations have elected to play the trading
game according to the new high-technology rules, participants have had
to run twice as fast in order to secure the rewards of leadership—
substantial export and foreign subsidiary sales, supra-normal profit margins, and for employees, compensation providing the purchasing power
to enjoy a high and improving standard of living.

For the United States, the transition to this new milieu has been particularly traumatic. In the first fifteen years following World War II, its
manufacturing enterprises enjoyed almost unchallenged supremacy
across a broad spectrum of high-technology industries. During the past
three decades, however, few have escaped tough technological challenges from abroad. That these challenges have been met with less than
overwhelming success is evidenced by rapidly rising imports of high-
technology merchandise, a steep decline in the U.S. dollar's overseas
purchasing power, stagnant productivity growth, and commensurately
disappointing real income gains for most U.S. workers. The shortfall of
means relative to individual aspirations has in turn made it difficult to
muster political support for investments in improved education, health,
and physical infrastructure—investments important if a high rate of economic growth is to be sustained. To be sure, the United States remains
a giant in the world arena. But the signs of *relative* economic decline

abound. And unless the direction of change is reversed, the United States risks falling prey to a debilitating case of the "British sickness."

Company Responses to High-Technology Challenges

Even in their "new" versions, the theories of international trade and comparative advantage paint with broad brush strokes. They are more "macro" than "micro." This book deviates from convention, adopting a "micro-micro" perspective. Its central thesis is that comparative advantage in the production of high-technology merchandise is won in the R&D laboratories and factory trenches. A key complementary role is played by the field generals who decide what new industrial products and processes to develop, how quickly to develop them, and (especially) how to respond to new technological challenges from determined overseas opponents.

Our case studies and statistical analyses have tried to determine how aggressive U.S. firms' responses to these challenges have been and to identify the reasons for diverse reaction patterns.

On the "how aggressively" question, the case studies, which deliberately avoided well-known failures such as steel and automobiles, provide a view inclining more toward optimism than pessimism. In wet shavers, color film, medical imaging apparatus, fiber optics, and earth-moving equipment, U.S.-based companies reacted vigorously and defended their traditional markets with considerable success. In airliners, Boeing fought hard but was forced to cede a sizable market share to its subsidized overseas rival, Airbus Industrie. After a slow start, the tire companies mounted a concerted technological defense, but the battle ended with much of U.S. production in the hands of foreign-owned enterprises. In central office digital switches, AT&T settled for a roughly equal division of market shares with challenger Northern Telecom. U.S. pocket calculator makers were routed in the high-volume, low-cost market segment but retained leadership in programmable devices. In three other product lines—color television sets, VCRs, and facsimile machines—U.S. firms' reactions were unambiguously submissive, and defeat was virtually total.

Our statistical investigation of companies' R&D reactions to import competition yields a somewhat more pessimistic picture. On average, the short-run R&D reaction to increasing imports was submissive—R&D/ sales ratios fell, albeit modestly. Our analysis of 17-year growth patterns suggests that companies' longer-run reactions edged toward the aggressive side, but the tendency was too unsystematic to be statistically significant. The long-run growth analysis implies that an average company, with served industry import shares rising by 0.64 percentage points per year, responded with R&D/sales ratios increasing at an average annual rate of 0.20 percent per year. Compounded over the 17-year sample span, this works out to cumulative growth of about 3 percent—for instance, with R&D moving from 3.15 to 3.25 percent of sales on average.

There is, to be sure, danger in putting too much stress on the R&D numbers. "Working smarter" might be a more effective response to high-technology import challenges than allocating more funds to research and new product development. But our case studies provided little support for the notion that R&D efficiency gains are commonplace. Among the companies emphasized by the case studies, only Caterpillar appears to have accelerated the flow of new products to market while cutting R&D outlays.

It is at least equally important to understand why some companies reacted aggressively while others cut back their R&D efforts. Company and industry structure mattered in a number of ways. For one, multinational enterprises reacted more heterogeneously than firms whose R&D activities were confined to the domestic economy. Indeed, companies with R&D operations outside the United States had on average a *zero* reaction to changes in net exports, whereas U.S. market specialists had strongly submissive reactions. Case study evidence suggested that R&D multinationals reacted differently because they were able to offset falling exports with imports from overseas subsidiaries, because they could change overseas as well as domestic R&D programs in response to altered strategic threats, and because their overseas operations provided a "distant early warning line" to detect new products and processes introduced first in foreign markets.[1]

Company reactions tended also to be more aggressive, the greater a company's domestic sales were, the more concentrated were the U.S. markets in which the company operated, and the more diversified the company's domestic operations were. The first two patterns almost surely reflect predictions from the pure theory of inter-firm R&D rivalry. Companies with large and (especially) dominant market positions have only weak incentives to be the first mover in a new product technology,

---

1. The analogy here is to the DEW line, a radar system constructed across Alaska and Canada during the 1950s to provide early warning of air attacks across the Arctic Ocean.

but when substantial sales are threatened by the incursion of "new kids on the block," the incumbents are inclined toward strongly aggressive "fast second" responses. Supporting this inference is the evidence that import penetration rose more rapidly in highly concentrated industries. Thus, powerful domestic firms conducted their business in a way that attracted high-technology imports, but when the threat escalated, they fought back tenaciously.

The diversification result is more difficult to interpret. It may reflect greater shock absorption capacity or the ability to cross-subsidize lines under attack from imports. However, our case study of the consumer electronics industry is inconsistent with this view. RCA did not meet the challenge of improved color television sets and new VCRs from Japan because, tapping television profits in an unsuccessful attempt to diversify into the computer business, it treated television set manufacturing as a "cash cow." When the diversification effort failed, traditional lines were hurt too. The tension here between case study and statistical evidence is by no means irreconcilable. There is sufficient "noise" in the statistical results that deviant cases should elicit no surprise.

To an increasing extent during the 1970s and 1980s, the United States government provided domestic industries special protection from import competition in the form of voluntary restraint agreements, mandatory quotas, and elevated tariffs. Our statistical analysis revealed that the more heavily protected a company's lines of business were, the more likely submissive R&D reactions to import inroads were. Conceivably, this could be interpreted to reflect a chain of causation in which import penetration engendered industry distress, which in turn led to reduced R&D, and the erection of trade barriers failed to alleviate the R&D-reducing distress. However, because the result persists when additional variables controlling for current and long-term changes in import penetration and profitability are included, it seems more plausible to conclude that having been granted protection, domestic producers rested on their technological oars. To the extent that protection schemes are intended to support and stimulate industry recuperation and not merely to alleviate pain, a significant policy shortcoming is implied.

In a much-debated 1980 article, Robert Hayes and William Abernathy argued that by emphasizing financial manipulation and "managing by the numbers" rather than "hands on mastery" of relevant technologies, American managers were undermining the long-run strength of the enterprises under their stewardship.[2] Our research provides some support for their thesis. Among the relatively technology-oriented companies in our statistical sample, the fraction in which at least one of the top two executive officers had an educational background in science or engineering peaked at 71.5 percent in 1980 and then declined five percentage points by 1987. In the analysis of import responses, companies headed by technically educated individuals were found to react more aggressively to rising imports, all else equal.

Other hypotheses were tested with mixed results. Companies with higher profit margins were found to support R&D more intensively and to raise their R&D/sales ratios more between 1971 and 1987, although the short-run impact of increased R&D was a decline in profit margins. There was weak evidence of more aggressive reactions when racing down learning curves was an important means of appropriating the benefits from new technology. Perhaps the most surprising finding was the consistent tendency toward submissive reactions in industries to which basic academic research in physics, chemistry, and biology was particularly relevant. A possible explanation is that basic knowledge diffuses rapidly to foreign rivals as well as to domestic firms, and Japanese companies, which have mounted the most pervasive technological challenge to U.S. enterprises, have been especially adept at exploiting the innovation possibilities opened up by advances in academic science.

Implications for World Trading Positions

The wide diversity of company responses emerged as a major theme from both our case studies and the statistical analyses. This apparent randomness of corporate behavior could have important implications for the future of American industry. If in an appreciable fraction of cases U.S. companies react submissively to high-technology challenges from abroad, and if in a larger plurality they maintain R&D "business as usual," they must in the long run cede substantial segments of the industrial technology frontier to aggressive foreign rivals. During the 1970s and 1980s, when technological challenges came with increasing frequency, many leadership positions were lost through the behavioral responses

---

2. Robert H. Hayes and William J. Abernathy, "Managing Our Way to Economic Decline," *Harvard Business Review*, July–August 1980, pp. 67–77.

characterized by our analyses. That the international competition will be at least as tough in the future seems certain. If aggressive responses continue to be more the exception than the rule, the erosion of U.S. technological leadership will almost surely continue.

This could among other things mean a change in the structure of U.S. technological efforts. Historically, Daniele Archibugi and Mario Pianta have shown,[3] there has been a strong positive correlation between the absolute size of national R&D efforts and the breadth of those efforts—the latter measured by the extent to which the patents received by domestic companies, or citations to those patents, covered the entire spectrum of industrial applications. As the nation with the largest R&D effort, the United States also encompassed the most widely diversified array of technological bets. Smaller nations targeted their R&D efforts on narrower industrial niches—presumably, those in which they had the best chances of reaching or holding the frontier. There is evidence of increasing specialization among both small and large nations with the passage of time.

As technological challenges from other nations persist, U.S. companies will have to choose. If they continue collectively to cover the whole array of industrial technologies, their efforts are likely to be outmatched by deeper, more narrowly focused programs overseas—barring a substantial increase in the total resources U.S. industry devotes to technological innovation. The alternative is to back away from some fields and to become more specialized, like smaller nations but on a broader plane. Gaps in the domestic coverage of product characteristics will be filled by importing: to each its niche, letting intra-industry trade ensure that specialized domestic demands are satisfied. If this happens, success will depend upon making good choices regarding where to withdraw and where to concentrate one's forces.

R&D Reactions and the Invisible Hand

In this respect, company reactions to international high-technology competition play a key role in the dynamic counterpart of Adam Smith's

---

3. Daniele Archibugi and Mario Pianta, "Specialization and Size of Technological Activities in Industrial Countries: The Analysis of Patent Data," in Mark Perlman and F. M. Scherer, eds., *Entrepreneurship, Technological Innovation, and Economic Growth* (Ann Arbor: University of Michigan Press, 1992).

"invisible hand." Aggressive reactions are the microeconomic mechanism through which positions of international trading dominance are defended and held. Submissive reactions are the mechanism through which comparative advantage is ceded to other nations' enterprises.

When tradable goods flow freely through world markets and the invisible hand nominates national industries to enjoy comparative advantage in the goods for which they have made key product or process technology contributions, substantial benefits accrue. In the aggregate, dynamic free trade is a positive-sum game. But some gain more than others, and there can be net losers. The division of benefits depends inter alia upon the terms of trade, and as Alfred Marshall recognized a century ago, the terms of trade are likely to be skewed in favor of nations with strong positions in technically advanced products.[4]

Recognizing that technological leadership affects the terms of trade and hence the distribution of benefits from trade, many nations attempt by means of subsidies and the exclusion of high-technology imports to propel their national champions to leadership positions, thereby interfering with the invisible hand's functioning. There is reason to believe that the combined value of the world trading game's payoffs declines as a result of such interventions.

Submissive reactions to rival R&D efforts are triggered, the relevant theory reveals, when the rival enjoys either a substantial time lead or lower R&D costs. Government R&D subsidies are one way a participant in an R&D "race" can gain an advantage over its rivals and induce them to submit. If the unsubsidized and submitting rivals would otherwise maintain comparative advantage, the dynamic allocation of resources is distorted. If the asymmetry of costs is not large, subsidies may merely spur the acceleration of R&D efforts, leading to R&D costs higher than they would be in the absence of artificial stimulation and, less certainly, to wasteful duplication of R&D efforts.

Whether intensified international R&D rivalry, stimulated by national subsidies or not, leads to wasteful duplication depends upon complex circumstances. If as a result of R&D rivalry firms move to niche-filling strategies, intra-industry trade in technically differentiated products will grow. Our statistical analyses point strongly to high R&D intensity as a basis for extensive intra-industry trade. From the relevant theory, we know that such niche-filling can be either welfare-enhancing or welfare-

---

4. See Chapter 1, note 15 supra.

reducing. On the one hand, the more it is driven mainly by the desire to "cannibalize" substantial surpluses from substitute products offered by other producers, the more likely it is that wasteful duplication will result from monopolistic competition in international trade. On the other hand, the more the new products offered in this way satisfy unique and previously unmet needs, the less waste and the more genuine social benefit such monopolistic competition is likely to generate.

Also, the development of new products and production processes often takes considerable time and large lumps of resources that, once invested, are sunk costs. This lumpiness makes it difficult for R&D resource allocation processes to achieve the step-by-step convergence to equilibrium expected when normally functioning "static" markets are subjected to shocks. When several symmetrically positioned rivals struggle for advantage in a promising new technology, large and possibly duplicative investments may be sunk before some subset of winners emerges. The resource costs are even larger when learning-by-doing economies are important, whether the rivals fight a price war in the struggle to race down learning curves, as in electronic calculators and television sets, or passively divide up the resulting market among themselves.[5] Still, it must be recognized again that "duplication" can yield benefits as well as costs. In particular, the more uncertain R&D outcomes are, the more desirable it is to have several firms, and arguably firms reflecting diverse national traditions and points of view, vying in parallel for an early success—especially when success will confer substantial non-cannibalized benefits upon both producers and consumers.[6]

These are variables about which we have little systematic quantitative knowledge. The most that can be said is that the "invisible hand" guiding research and development resource allocation decisions under conditions of international rivalry does not operate flawlessly. It fails, perhaps frequently, especially when its operation is distorted by governmental subsidies and restrictions. Its continued acceptance has a simple rationale:

---

5. The more aggressive of these strategies tends to be more economical in the long run, especially when product generation lives are relatively short. See David Ross, "Learning to Dominate," *Journal of Industrial Economics,* 34 (June 1986), pp. 337–354.

6. See Richard R. Nelson, "Uncertainty, Learning, and the Economics of Parallel Research and Development Efforts," *Review of Economics and Statistics,* 43 (November 1961), pp. 351–368; F. M. Scherer, "Time-Cost Tradeoffs in Uncertain Empirical Research Projects," *Naval Research Logistics Quarterly,* 13 (March 1966), pp. 71–82; and Burton H. Klein, *Dynamic Economics* (Cambridge: Harvard University Press, 1977).

it is difficult to find superior resource allocation mechanisms. Especially deficient in this respect, if the past history of individual national efforts to "steer" civilian sector R&D resource allocation is any guide, would be schemes in which pervasive governmental or inter-governmental planning and control are substituted for the more-or-less independent decision-making of technically adept firms.[7]

Policy Measures

How well U.S.-based enterprises perform in the action–reaction process of international high-technology competition matters for the welfare of Americans. Given the substantial capabilities and aspirations of other nations, it would be unrealistic to expect U.S. companies to dominate high-technology trade to the extent they did during the 1950s and 1960s. More dispersed leadership and specialization are inevitable. Yet as the largest trading partner among the technologically proficient nations, the United States should at least aspire to hold a plurality of leadership positions. Success in that endeavor will bring favorable terms of trade, supra-normal profits, and higher real incomes. Moreover, by excelling in technologies with wide-ranging applications, such as semiconductors, computers, superconductive materials, and genetic engineering, U.S. companies put themselves on a dynamic learning path that makes success in the development of future technologies more likely. Once leadership is lost to more aggressive rivals, it will be difficult to overcome know-how and experience handicaps in subsequent rounds.

Many of our case studies are success stories, documenting vigorous and effective responses by U.S. companies to new high-technology competition. Some are failure stories, and a few lie in the gray area. Whatever the mix of successes and failures, there is room for improvement.

Our research was meant to be descriptive, not prescriptive, and from it, policy recommendations for improving the technological performance of American industry do not leap out. Policy judgments must in any event rest upon a broader base of knowledge, analysis, and individual values. Recognizing these limitations, we proceed with caution, focusing on

---

7. See, for example, Richard R. Nelson, ed., *High-Technology Policies: A Five-Nation Comparison* (Washington, D.C.: American Enterprise Institute, 1984). The most important counter-example is Japan's considerable (but not uniform) success in targeting industries for technological stimulation. See Christopher Freeman, *Technology Policy and Economic Performance: Lessons from Japan* (London: Pinter, 1987).

problems on which our research has added new, even if incomplete, insights.

Management

The failure of some U.S. businesses to meet technical challenges from abroad has been first and foremost a managerial failure. At key points, the managers of U.S. companies did not allocate resources needed to remain technologically competitive, carried existing developments into the market too slowly, and failed to maintain an organizational environment conducive to innovation. Doing these things is not easy. But failure is more likely when top managers lack an appreciation for, and skill in, making innovations happen. Our statistical analyses suggest that R&D support is greater, and reactions to rising import competition are more aggressive, when top managers are educated in science or engineering. In view of this, it is disconcerting to find, at least for our sample of R&D-oriented corporations, a shift away from technically educated leaders during the 1980s. The boards of directors for corporations operating in technologically dynamic industries should ensure that appropriately skilled managers are groomed for top positions.

If one accepts as inevitable the trend toward leadership by executives with MBAs (for our company sample, increasing from 24 percent in 1971 to 42 percent in 1987), it is essential that management schools provide their students in-depth education on the significance of technological innovation, the challenges it poses, and means of sustaining it. A 1982 survey disclosed that at most, 11.5 percent of the MBA students in twelve leading business schools acquired systematic full-semester exposure to questions of managing technological innovation.[8] The situation may have improved slightly since then,[9] but unless the top business school faculties force a change, new generations of prospective American business leaders will be sent out with an inadequate understanding of how their companies can be kept internationally competitive.

---

8. F. M. Scherer, "Technological Change and the Modern Corporation," in Betty Bock, Harvey J. Goldschmid, Ira M. Millstein, and F. M. Scherer, eds., *The Impact of the Modern Corporation* (New York: Columbia University Press, 1984), p. 275.

9. In 1991, the American Assembly of Collegiate Schools of Business required that for program accreditation, all MBA candidates must receive "more" training in technology.

Science and Production Engineering

The television, VCR, and (for EMI) CT scanner case studies illuminate a related failure—the tendency of research and development organizations to overstress "ivory tower" prototype design while giving short shrift to more mundane production engineering and quality assurance. It simply will not do, as in RCA's failed videodisc effort, for product design staff to view production engineering as a remote lower-caste activity out in Indianapolis, "regular and extended trips" to which were "the hardest to bear" of tedious routines.[10] On this, we concur in the recommendations of the MIT Commission on Industrial Productivity.[11] Business firms must revamp their training programs to bring R&D staff into early and continuing contact with production operations, and their reward systems must be restructured to ensure that cost-conscious, reliable manufacturing is encouraged on a par with innovative design. But reform must commence at an earlier stage, in the universities, with solid hands-on training for engineering and applied science students in production methods and ways of improving them.

The role of academic science must also not be forgotten. A surprise in our statistical analyses was the discovery that companies reacted more submissively in fields where advances in the basic sciences were most relevant. Evidently, U.S. manufacturers (or at least, the relatively well-established companies in our sample) have not been sufficiently energetic in tapping the ideas and opportunities basic research generates, both in the United States and abroad. What seems needed is more a change in company and university attitudes than in public policies. There are already substantial tax code incentives for grants and gifts that help build company ties to university research laboratories. Companies should encourage their R&D staff to cultivate those ties and keep abreast of new discoveries, and university administrators and opinion leaders should foster cooperation by their faculty in information exchanges with industry. As the locus of path-breaking scientific discovery becomes increasingly dispersed throughout the world, U.S. scientists and engineers must rebuild communication networks with leading scientific investigators over-

---

10. Margaret B. W. Graham, *RCA and the VideoDisc* (Cambridge: Cambridge University Press, 1986), p. 180. See also pp. 152–161, 186, and 200–201.

11. Michael L. Dertouzos, Richard K. Lester, and Robert M. Solow, *Made in America: Regaining the Productive Edge* (Cambridge: MIT Press, 1989), chaps. 10–12.

seas. Such networks flourished during the 1920s and 1930s, but atrophied as the United States came to dominate international science.

Industrial Structure

Industrial structure occupied a prominent position at several key points in our analysis. The policy implications, however, are complex, requiring complementary evidence.

The reactions of multinational corporations to net export declines, although close to zero on average, were at least more aggressive than those of firms performing R&D only in the United States. From a broader perspective, there are few disadvantages to the extension of overseas operations by companies at home in the United States,[12] and the ability of multinationals to detect new overseas technological threats in their infancy and take appropriate defensive measures is a significant advantage. It is far from clear, however, that the benefits from multinationality can be enhanced through active policy initiatives. Encouraging firms to commence or expand overseas operations when they lack appropriate managerial skills and intangible assets on which to build is a likely prescription for failure. If outbound foreign investment does not emanate naturally from certain domestic industries, it would be better for the government to support low-cost alternatives to multinational firms' information networks. In particular, the "distant early warning" function could be augmented by having government agencies such as the Office of Technology Assessment and the National Institute of Standards and Technology intensify their surveillance of world scientific and technological horizons to identify new opportunities and threats relevant to domestic industries. An important part of the effort would be an active program for pinpointing U.S. industries that could benefit from newly analyzed information and ensuring that the information is disseminated.

Our statistical analysis revealed that large, diversified firms at home in concentrated industries reacted more aggressively to high-technology import shocks. This would appear to support "Schumpeterian" argu-

---

12. See, for instance, Raymond Vernon, *Sovereignty at Bay* (New York: Basic Books, 1971), chap. 5; C. Fred Bergsten, Thomas Horst, and Theodore Morgan, *American Multinationals and American Interests* (Washington, D.C.: Brookings, 1978); Richard E. Caves, *Multinational Enterprise and Economic Analysis* (Cambridge: Cambridge University Press, 1982); and Edward M. Graham and Paul R. Krugman, *Foreign Direct Investment in the United States* (Washington, D.C.: Institute for International Economics, 1989), chap. 3.

ments that monopoly power and large enterprise size are conducive to technological innovation.[13] However, inferences must be drawn cautiously here. High seller concentration was associated not only with more aggressive reactions to a given import threat but also with more rapid import penetration. The most likely explanation is that large companies in concentrated industries have pursued policies that let overseas rivals jump ahead technically and penetrate the U.S. market with their new or lower-cost products, but then fought back with classic "fast second" responses. This interpretation is consistent with the case study evidence on shaving gear, radial tires, color film, CT scanners, and digital switches. Given the imprecision of our statistical estimates, it is difficult to be certain whether domestic markets would have been less vulnerable had a different, less concentrated, market structure existed.

Wherever the truth lies on this point, there are possible implications for antitrust policy.

During the 1980s, the Reagan Administration urged unsuccessfully that the U.S. merger laws be amended to allow domestic enterprises to restructure and attain the larger scales believed necessary inter alia to meet growing foreign competition.[14] Our Chapter 5 analysis provides some support—equivocal, to be sure—for that position. However, other evidence must also be weighed. The preponderantly conglomerate merger wave that occurred during the 1960s and early 1970s is now considered to have worked more harm than good.[15] Among other things, there is evidence that R&D/sales ratios were reduced, not increased, in the wake of conglomerate acquisitions.[16] Although horizontal merger activity escalated in the United States during the 1980s, there is little systematic evidence on how it affected R&D and innovation. The experience from Europe with large-scale horizontal mergers, often promoted by national

---

13. For a review of the arguments, see F. M. Scherer, "Schumpeter and Plausible Capitalism," *Journal of Economic Literature,* forthcoming.

14. See the testimony of Commerce Secretary Malcolm Baldrige in U.S. House of Representatives, Committee on Banking, Finance, and Urban Affairs, Subcommittee on Economic Stabilization, Hearings, *Structuring American Industry for Global Competition* (Washington, D.C.: U.S. GPO, 1986), pp. 5–42.

15. See David J. Ravenscraft and F. M. Scherer, *Mergers, Sell-offs, and Economic Efficiency* (Washington, D.C.: Brookings, 1987); Michael E. Porter, "From Competitive Advantage to Corporate Strategy," *Harvard Business Review,* May–June 1987, pp. 43–59; and Michael C. Jensen, "Eclipse of the Public Corporation," *Harvard Business Review,* September–October 1989, pp. 61–74.

16. Ravenscraft and Scherer, *Mergers,* pp. 120–122.

governments in the hope of securing scale economies and invigorating innovation efforts, shows that more often than not, those hopes were followed by disappointment.[17] Clearly, encouraging large firm sizes and concentration through merger is no magic path to international competitiveness.

In view of the conflicting evidence, a milder step is all that seems warranted. The merger laws should be amended to let would-be horizontal merger partners overcome the presumption of illegality associated with substantially increased concentration if they can show that significant improvements in operating efficiency and/or technological innovation can be realized as a consequence.[18] Because it is often difficult to prove the likelihood of efficiencies in advance of mergers, a two-stage procedure would be appropriate. If substantial but inconclusive pre-merger evidence of likely pro-competitive gains is presented, the merger would be approved for a "trial run" of several years. If the predicted gains have not in fact been achieved by the end of the trial period, the merger or an equivalent bundle of assets would be divested to restore the pre-merger competitive status quo.

It would be advantageous to incorporate explicitly a similar philosophy into the laws governing monopolization, notably, Section 2 of the Sherman Act.[19] Companies charged with monopolization should be able to

---

17. See Dennis Mueller et al., *The Determinants and Effects of Mergers: An International Comparison* (Cambridge, Mass.: Oelgeschlager, Gunn, & Hain, 1980); Keith Cowling, Paul Stoneman, and John Cubbin, *Mergers and Economic Performance* (Cambridge: Cambridge University Press, 1980); Gerald D. Newbould, *Management and Merger Activity* (Liverpool: Guthstead, 1970); Paul Geroski and Alexis Jacquemin, "Large Firms in the European Corporate Economy and Industrial Policy in the 1980s," in Jacquemin, ed., *European Industry: Public Policy and Corporate Strategy* (Oxford: Clarendon Press, 1984), pp. 344–349; and William James Adams, *Restructuring the French Economy* (Washington, D.C.: Brookings, 1989), chap. 6.

18. A pioneering proposal along these lines was Oliver E. Williamson, "Economies as an Antitrust Defense: The Welfare Tradeoffs," *American Economic Review,* 58 (March 1968), pp. 18–36. For elaboration, see F. M. Scherer and David Ross, *Industrial Market Structure and Economic Performance* (third ed.; Boston: Houghton-Mifflin, 1990), pp. 186–188.

19. The case law has moved erratically in this direction during the past two decades. See, in particular, the Federal Trade Commission decision *In the matter of E. I. du Pont de Nemours & Co.* (the titanium dioxide case), 96 F.T.C. 653 (1980). AT&T placed considerable emphasis upon its innovativeness in defending itself against the major structural monopolization action settled by negotiation in 1982. Our case studies reveal that innovation in digital central office switches and facsimile machines was accelerated when AT&T's monopoly positions were undermined, so it is questionable whether an innovation defense could, or should, have succeeded.

defend themselves against structural breakup by showing that their dominant market positions were based primarily upon superior technological innovation, consistently pursued. This defense would be accorded even greater weight if the respondent company could be shown to have sustained strong export and/or foreign direct investment performance.

Most U.S. monopolization cases have been settled, not through structural reorganization, but through the application of more narrowly targeted remedies. One of the most common instruments, particularly during the 1940s and 1950s, was the compulsory licensing of patents.[20] Most of these compulsory licensing decrees were issued before high-technology import challenges to U.S. industries multiplied. The experience of RCA with its color television patents and Xerox with its copier patents[21] advises greater caution in future cases to ensure that opening up patent portfolios does not cause a loss of U.S. firms' advantages vis-à-vis overseas rivals.

Trade Policy

Finally, our research has implications for international trade policy, as adjudicated by the GATT authorities in Geneva and several agencies within the United States—the President's Trade Representative, the International Trade Commission, and the Commerce Department's International Trade Administration.

Massive, narrowly targeted subsidies such as those to Airbus Industrie plainly distort dynamic comparative advantage. They should be combatted, despite the embarrassment that comes when those who cast the first stone sin on other fronts. The invisible hand is also led astray by the even more ubiquitous exercise of preference for home-developed equipment in procurement by nationalized and quasi-public enterprises and, in Japan, by companies affiliated in *Keiretsu* groups.[22] Among the products covered by our case studies, central office digital switches and fiber optic cables

---

20. See F. M. Scherer, *The Economic Effects of Compulsory Patent Licensing* (New York: New York University Monograph Series in Finance and Economics, 1977).

21. On Xerox, see Timothy F. Bresnahan, "Post-Entry Competition in the Plain Paper Copier Market," *American Economic Review,* 75 (May 1985), pp. 15–19. Rapidly increasing imports during the mid-1970s forced Xerox to improve its machine designs and reduce costs. But those reactions did not reverse the tide; they only stemmed it. Imports continue to dominate much of the medium-volume market.

22. See Robert Z. Lawrence, "Efficient or Exclusivist? The Import Behavior of Japanese Corporate Groups," *Brookings Papers on Economic Activity,* 1991, no. 1, pp. 311–330.

provide prominent examples. The United States has attempted to reverse such practices by threatening retaliatory restrictions under the "Super 301" clause of the amended U.S. Trade Act. Such threats cause considerable resentment and could, if carried out on the broad plane the U.S. Congress apparently intended, induce escalating counter-retaliation and a breakdown of mutually advantageous trading relations. It would be far better to deal with the problem through a framework of internationally accepted rules. At present, GATT's provisions do not effectively reach domestic preference behavior. Filling that lacuna should be a focal point in future negotiations to extend GATT's reach.

Although trade restraints serve a valid role in disciplining export practices that break the rules of the trading game, they can also stifle the competitive pressures that stimulate lagging industries. Our statistical analysis shows that both in the short run and the long run, companies' R&D spending responded less aggressively on average to import shocks of given magnitude when trade barriers were in place. In automobiles, steel, television receivers, large electric motors, motorcycles, and other industries, trade barriers have been erected to protect companies in trouble because of their own technological failures. Granted, once U.S. firms have fallen behind, it may be difficult for them to recoup unless they are sheltered to some extent from the competition of technically superior overseas rivals. To avoid the problems identified by our research, a deft combination of carrots and sticks is recommended. Specifically, the U.S. Trade Act should be amended to provide that if technological laggardliness or production inefficiency is found to be a significant contributor to the competitive injury experienced by U.S. firms as a consequence of imports, protection should be extended for not more than five years. Equally important, the amount of protection should decline automatically over time—for example, as in the 1983 Harley-Davidson Section 201 case, with tariff increments of 45 percent in the first year, 35 percent in the second year, 20 percent in the third year, 15 percent in the fourth year, and dropping to 10 percent in the last year.[23] In this way, protected enterprises will recognize that unless they respond aggressively with improved products and production processes, they will be neither internationally competitive nor protected in the foreseeable future.

---

23. See "How Harley Beat Back the Japanese," *Fortune,* September 25, 1989, pp. 155–164; and "How Harley Outfoxed Japan with Exports," *New York Times,* August 12, 1990, p. F5.

Future Research

The research reported here is a first step into a set of relationships that has previously received little systematic attention from scholars. It is by no means the last word on the subject. Much remains to be done.

Time and data limitations required that our statistical analysis end with the year 1987. There is reason to believe that company reactions to high-technology import competition have been changing over time. During the 1970s, many companies were caught by surprise by the new competition from abroad. Consistent with the asymmetries that often accompany surprise, R&D reactions appear to have been submissive on average. As it became clear that the new competition was not merely transitory, companies realized that they had to redouble their technological efforts to remain competitive, and reactions may have become more aggressive. During the late 1980s, U.S. industrial R&D spending growth again stagnated. It is unclear whether this new trend was influenced by the evolving pattern of international competition. The falling value of the dollar may have blunted the import threat, leading U.S. companies once again to relax their R&D efforts, or some companies may have found the new challenges to be so enduring and formidable that submissive reactions again predominated. Continuing research is needed to illuminate the changing behavioral patterns in all their subtlety.

To a greatly increased degree during the 1980s, foreign firms that once competed mainly by importing have established substantial operations within the United States. The data readily available to us permitted only a cursory statistical analysis of how foreign direct investment in the United States is related to import activity. Even less could be done on whether FDI has asymmetric incentive effects on U.S.-based companies' research and development spending. With much greater effort, data on foreign ownership could be assembled at the four-digit level of detail used in our import analyses. We commend the task to future investigators.

Author Index

Subject Index